Tourism Dynamics: New perspectives and changing directions

Nikolaos Pappas and Anna Farmaki

(G) Goodfellow Publishers Ltd

(G) Published by Goodfellow Publishers Limited,
26 Home Close, Wolvercote, Oxford OX2 8PS
http://www.goodfellowpublishers.com

British Library Cataloguing in Publication Data: a catalogue record for this title is available from the British Library.

Library of Congress Catalog Card Number: on file.

ISBN: 978-1-911635-93-2

DOI: 10.23912/9781911635932-4921

Design and typesetting by P.K. McBride, www.macbride.org.uk

Cover design by Cylinder

Contents

Contributors

Editors

Nikolaos Pappas is an Associate Professor in Tourism, Hospitality and Events, and the Director of CERTE (Centre for Research in Tourism Excellence) at the University of Sunderland, UK. He holds a doctorate (PhD) in Tourism Planning and Development, and a post-doctorate (PDoc) in Risk and Crisis Management. He started his career in the tourism and hospitality industry in 1990, and for 10 years (2001-2010) he was also engaged in enterprising consultancy. Since 2001, he has been an academic in higher education, gaining experience from several institutions in Greece (Technological Education Institute of Crete, Hellenic Open University, EKDDA) and the UK (Derby, Northampton, Leeds Beckett, UWL).

Nikolaos has participated in nationally and internationally funded research projects since 1998. He has numerous publications in international scientific refereed journals and conferences, and he is a reviewer in several academic journals including *Annals of Tourism Research, Journal of Sustainable Tourism, International Journal of Contemporary Hospitality Management, International Journal of Hospitality Management, Journal of Business Research,* and *Tourism Management*. His academic interests include crisis management communications, tourism and hospitality management, e-tourism and e-marketing, and destination planning and development.

Anna Farmaki is Assistant Professor in Tourism Management at the Cyprus University of Technology, Cyprus. She holds a doctorate (PhD) in marketing. She has worked for several years in private institutions of tertiary education in Cyprus and was course leader of the undergraduate hospitality and tourism management programme at the University of Central Lancashire (Cyprus). She has published extensively in reputable, peer-reviewed academic journals and has presented her work in various international conferences, seminars and workshops.

Anna is a Fellow of the Higher Education Academy (UK) and a member of the Management Committee in several COST Actions. She has also been involved and/or led several nationally and internationally funded research projects. In addition, she is a member of the Editorial Board of the *International Journal of Contemporary Hospitality Management,*

Tourism Management Perspectives, Tourism Review and the *Journal of Service Theory and Practice.* She authored the book *Tourism and Society* published in Greek by Ion Publishing. Her research interests lie primarily in the areas of tourism planning and development with emphasis on sustainable tourism, and tourist behaviour.

Authors

Pavlos Arvanitis is Senior Lecturer in Aviation Management, University of Bedfordshire, UK. He has over 20 years' research experience in the fields of air transport and tourism and over 10 years' teaching experience in Higher Education. His professional career started as a country sales executive in Greece, for a small regional Italian carrier based in Venice.

Iride Azara is Senior Lecturer in Tourism and Events at the University of Derby and an active researcher across national and international research groups and associations. Her research focuses on cultural and heritage tourism experiences as resources for the sustainable revitalisation of communities and destinations alike.

David Brown is Leader of the PhD Programme at Newcastle Business School, Northumbria University. His research interests include the interplay of landscape, exercise, mobility and self within serious leisure and adventure tourism – especially within hillwalking and long-distance walking. He has also published research on qualitative methodology, pedagogy, and aspects of marketing and sustainability.

Christina Dimitriou is a Certified Hospitality Educator (CHE) and Certified Online Instructor who has been teaching undergraduate and graduate tourism and hospitality courses for over 20 years in Europe and the United States. Research accomplishments include journal articles, book chapters, conference posters and presentations at regional, national and international conferences.

Nilgün Handan Eskitark is a senior student in the Tourism Administration Department at Boğaziçi University, in Istanbul. She has held internships in eco-labeled accommodation businesses and for four years ran a cafe serving traditional dishes in a folkloric concept. She has been interested in understanding thin rural areas and their cultures. So, she spent a lot time on cycling around Turkey's and neighbourhood countries' rural destinations to investigate the lifestyle of villagers. Her aim is to pursue a Master's degree and a PhD following her research interests in sustainable destination development, as well as rural, cultural and eco-tourism.

Alba Font Barnet is a PhD candidate at the Research Group for Territorial Analysis and Tourist Studies, Department of Geography, Rovira i Virgili University. With research interest in the well-being of natural areas, she participates in projects of tourist experiences co-creation in Spain and Latin America Protected Areas.

Carla Fraga is an associate professor in the Department of Tourism and Heritage of the Universidade Federal do Estado do Rio de Janeiro, and holds a D.Sc. in Transport Engineering from the Programa de Engenharia de Transportes of the Universidade Federal do Rio de Janeiro. Her research interests are tourism, transportation and neurosciences.

Antonios Giannopoulos is an adjunct lecturer at the Department of Organisation Management, Marketing & Tourism, International Hellenic University and assistant professor at Deree – The American College of Greece. He holds a PhD in Marketing & Communication from the Athens University of Economics and Business. Following a 15-year academic and professional experience in marketing, he has presented research in services, tourism marketing, cultural diplomacy at international conferences (e.g. EMAC, CHRIE, TTRA) and has a portfolio of published scholarly articles (e.g. *Annals of Tourism Research, Managing Service Quality, Journal of Hospitality Marketing & Management*).

Elena Jelača is Master's graduate in Tourism Management from the University of Derby, UK. Her research interest is focused on Generation Y as backpackers and their attitudes and behaviours towards volunteering. She is passionate about understanding young traveller's motivations for backpacking, sustainability, and future employability in the context of volunteering.

Christina Karadimitriou is a PhD researcher at the University of Patras, Greece and the Public Relations Manager in ViRA (Virtual Reality Applications) in Athens. Her research interests include crisis management communications in tourism and hospitality, destination marketing, e-tourism and travel and tourism for disabled people.

Stella Kladou is a postdoctoral researcher at the Hellenic Mediterranean University. She has served as an Assistant Professor in the Tourism Administration Department at Boğaziçi University in Istanbul, Turkey and a Senior Lecturer at Sheffield Hallam University in the UK. She has published in prominent academic journals and books of international publishers, and presented at well-esteemed international conferences. She is an International Place Brand Specialist for the Place Brand Observer and in the scientific and editorial board of international conferences and the *Journal of Qualitative Research in Tourism* respectively. Her research interests include place branding, consumer behavior, cultural tourism, and digital marketing.

Elena Konovalov is currently an adjunct at the School of Business and Law at CQUniversity. Previously she worked there as a lecturer in Marketing and Management. Her doctoral research was focused on understanding community well-being and its links to tourism. Dr Konovalov's current research is focused on tourism marketing and green tourism practices.

Eleni Michopoulou is an associate professor in Business Management at the University of Derby, UK. Her research interests include technology acceptance and online behaviours, and accessible and wellness tourism marketing and management. She has published over 60 academic journal articles, book chapters and conference papers in these areas.

Gianna Moscardo is a professor in the School of Business at James Cook University. Her background in sociology supports her research interests in the areas of community responses to tourism and her background in psychology supports studies into tourist experiences and behaviours.

Laurie Murphy is an associate professor of tourism in the School of Business at James Cook University. Dr Murphy's research interests focus on improving tourism's contribution to regional communities with an emphasis on tourism marketing, including a focus on the backpacker market, destination image and choice and more recently destination branding and tourist shopping villages.

Marta Gemma Nel-lo Andreu is a professor in the Department of Geography at Rovira i Virgili University, and a member of the Research Group for Territorial Analysis and Tourist Studies. She is the coordinator of the International Network of Tourism and Development Cooperation Researchers.

Sylwia Orzeł is a master's degree student of Tourism and Recreation at the Institute of Geography, Pedagogical University in Krakow. She conducts her research in the area of the social impact of overtourism in tourism cities.

Andreas Papatheodorou is a prolific academic researcher and advisor in air transport and tourism economics. He is currently a professor in Industrial and Spatial Economics with Emphasis on Tourism at the University of the Aegean, Greece and an adjunct professor at the School of Aviation, University of New South Wales, Australia.

Jakub Para is a master's degree student of Tourism and Recreation at the Institute of Geography, Pedagogical University in Krakow. His research interest is the development of winter sports destinations in Poland.

Maria Psimouli holds a PhD in Marketing from the Athens University of Economics and Business. She is currently an adjunct professor at The American College of Greece, and she has previously worked as a research fellow at the Athens Laboratory of Research in Marketing and as a marketing executive in national and multinational companies. Her research interests lie in the areas of international marketing and place branding.

Airton Rodrigues is General Director of NENC (Núcleo de Estudos em Neurociência do Consumidor – Center for Studies in Consumer Neuroscience), and has a Doctorate in Cognitive Psychology (São Paulo University), Master in Psychology (São Paulo University), and MBA in Marketing (São Paulo University). He conducts experiments about the consumer's experience, in business segments, using different neuroscience methodologies.

George Skourtis is a Senior Lecturer at Christ Church Business School, Canterbury Christ Church University. He has published articles in international peer-reviewed journals and international conferences. His research interests include marketing theory, services marketing, tourism marketing and consumer behaviour. He holds a PhD in Management Science from Toulouse School of Management, France. He has conducted collaborative research with a number of academic institutions, including the University of Vienna, Wageningen UR, Dokuz Eylul University, Athens University of Economics and Business, University of Piraeus, Istanbul Technical University, Piraeus University of Applied Sciences.

Maria Zoi Spanaki is a PhD researcher in New Services Risk Managment and Academic Tutor in Strategies and Operations at the University of Sunderland. Working as an assistant guest services manager in the hotel industry, she is passionate about conducting research and gaining further knowledge related to business and tourism. Research interests: risk management perspectives and tourism innovation.

Adam R. Szromek is a professor of Social Science at the Silesian University of Technology in Poland. His research interests focus on heritage tourism and health tourism with special attention to spa tourism management and business models in spa enterprises. Prof. Szromek's research has appeared in many reputable scientific journals.

Sharon Wilson is interested in tourism mobilities, social sciences and the creative industries. Adopting interdisciplinary and experimental research approaches to the study of social phenomena, she adopts experimental methodologies that seek to understand human mobility in an imaginative and embodied ways. She is also the founder of the Mobilities Futures Research Network (https://mobilityfrn.com/)

Erisher Woyo is an MBA Programme Coordinator and Senior Lecturer at the Namibia Business School, University of Namibia. His ongoing work focuses on ICT and higher education, and tourism in distressed contexts. His work has been published in *Anatolia, Journal for Marketing of Higher Education, Development Southern Africa,* and several other outlets.

Michał Żemła is an associate professor of economics working at the Department of Tourism and Regional Studies at Pedagogical University in Krakow, Poland. His recent research interest includes the development of winter sports destinations, sharing economy in tourism, and overtourism.

Introduction

The travel, tourism and hospitality fields have seen dramatic changes during the last four decades (Gursoy & Sandstrom, 2016). These fields are considered as fairly young academic disciplines, whilst their brief history has ended up in a lack of consensus on their exposure and scope (Kandampully et al., 2014). According to Min et al. (2016), tourism and hospitality have been recently identified as independent academic categories (Web of Science), showcasing the substantial progression of the disciplines. This is due to the widespread contribution of tourism in local societies and national economies through the provision of economic development, the creation of job opportunities, the inclusion of locals in community participation and the enhancement of participatory democracy, the increase of income, the improvement of life quality, the creation of infrastructure, and the provision of opportunities for the preservation of natural and built environment (Farmaki, 2015). However, the tourism and hospitality industries are susceptible to risks and crises (Pappas, 2018), and tourism planners need to carefully take under consideration any current and future dynamics that may result in considerable change and create new challenges and opportunities (Pappas, 2021).

The tourism literature extensively incorporates current and future trends, challenges and applications (indicatively please read Fatima & Elbanna, 2020; Kim et al., 2018; and Wut et al., 2021). However, extant literature has been less preoccupied with issues concerning the collective new perspectives, challenges, and directions of the respective field and the effects that this change might have not only on the tourism and hospitality industries but also on the global economy and society. Indeed, given that tourism is the third largest economic sector worldwide (Fatima & Elbanna, 2020), the challenges and new directions impacting the industry are likely to exert further influence on the economic and social fabric of countries at the regional, national and international levels (Pappas & Bregoli, 2016).

The chapter contributions in this edited book highlight and substantiate several facets of emerging insights and routes in the local, regional, national and international levels. They also showcase the importance of relevant new perspectives and directions for both destinations and enterprises as well as regional, national and international economic and social development of the tourism industry. The originality of the book lies in the fact that there is no other book/textbook/handbook that incorporates new perspectives and directions in such a rapidly changing environment in tourism and hospitality. This is further strengthened from the fact that the edited book collects high-quality chapters from esteemed scholars around the world, and encourages diversity under a global perspective.

It is crucial for destinations and tourism practitioners to fully encapsulate the importance of emerging trends and, thus, gain a better understanding of the rapidly shifting business environment they are operating in. In so doing, destination planners, policymakers and industry practitioners will be better prepared for change and to embrace pertinent opportunities whilst dealing with potential threats. On a similar note, tourism and hospitality researchers will need to acknowledge the collective effects of new perspectives and directions on the industry. Specifically, a reading of the *Tourism Insights: New perspectives and changing directions* can assist scholars to better comprehend the ever-changing dynamics influencing the industry; thus, allowing for a more holistic examination of tourism-related phenomena and for a greater appreciation of the shifting nature of the tourism industry as a whole. Such knowledge is also important for educational purposes as tourism and hospitality students need to be better prepared for their operational and managerial role in the industry by adopting critical thinking with regard to its changing dynamics. Accordingly, tourism and hospitality educators need to adapt their curriculum to reflect the influence of such direction on the industry (Farmaki, 2018).

Notwithstanding, there are few sources that critically discuss the effects of new perspectives and changing directions on the tourism and hospitality industry in a holistic manner. The aim of this book is, therefore, to offer a platform allowing the critical discussion of pertinent perspectives and directions and, as such, provide theoretical and practical insights that may be of value to destination planners, policymakers, industry practitioners, educators and students. The book will, specifically, attempt to offer a multi-disciplinary and inter-disciplinary approach to the study of tourism and examine emerging topics that are likely to change the future of the industry, either positively or negatively. Overall, the book contributes to the advancement of knowledge surrounding emergent controversies pertaining to tourism sustainability and innovation.

The edited book may be considered a must-read piece of work for diverse audiences, since its contents can provide substantial theoretical and practical knowledge to an array of stakeholders relevant to tourism and its sub-sectors (hospitality; aviation etc). Specifically, by focusing not only on emerging perspectives and directions in tourism but also by providing a comprehensive overview of the continuously and rapidly changing dynamics in the industry, the book aims to contribute to the scholarly debate on the current status and the future evolution of tourism. Hence, audiences can vary from students, educators, early career and experienced researchers focusing on multi and inter-disciplinary areas as well as industry practitioners, destination planners and policymakers.

References

Farmaki, A. (2015). Regional network governance and sustainable tourism. *Tourism Geographies, 17*(3), 385-407.

Farmaki, A. (2018). Tourism and hospitality internships: A prologue to career intentions?. *Journal of Hospitality, Leisure, Sport & Tourism Education, 23*, 50-58.

Fatima, T., & Elbanna, S. (2020). Balanced scorecard in the hospitality and tourism industry: Past, present and future. *International Journal of Hospitality Management*, 91, (in press).

Gursoy, D., & Sandstrom, J.K. (2016). An updated ranking of hospitality and tourism journals. *Journal of Hospitality and Tourism Research*, 40(1), 3-18.

Kandampully, J., Keating, B.W., Kim, B.P., Mattila, A.S., Solnet, D. (2014). Service research in the hospitality literature insights from a systematic review. *Cornell Hospitality Quarterly*, 55(3), 287-299.

Kim, C.S., Bai, B.H., Kim, P.B., & Chon, K. (2018). Review of reviews: A systematic analysis of review papers in the hospitality and tourism literature. *International Journal of Hospitality Management*, 70, 49-58.

Min, H., Park, J., & Kim, H.J. (2016). Common method bias in hospitality research: a critical review of literature and an empirical study. *International Journal of Hospitality Management*, 56, 126-135.

Pappas N., & Bregoli, I. (2016) *Global dynamics in travel, tourism and hospitality*. Hershey – Pennsylvania: IGI Global.

Pappas, N. (2018). Hotel decision-making during multiple crises: A chaordic perspective. *Tourism Management*, 68, 450-464.

Pappas, N. (2021). COVID19: Holiday Intentions during a Pandemic. *Tourism Management*, (in press).

Wut, T.M., Xu, J., & Wong, S. (2021). Crisis management research (1985–2020) in the hospitality and tourism industry: A review and research agenda. *Tourism Management*, 85, (in press).

Part 1: Travel and Tourism

Preface

This is the first part of the edited book and consists of five chapters. More specifically, the first chapter is written by Elena Jelaca, Iride Azara, and Eleni Michopoulou, and focuses on the generation Ys' attitudes and behaviour towards engaging in volunteer tourism whilst backpacking. Marta Gemma Nel-lo Andreu, and Alba Font Barnet co-author the second chapter and examine experiential tourism in relation to improved holistic well-being, while they provide examples of experiences in protected natural areas. The third chapter provides a look at the future of tourism and discusses consumer neuroscience in the context of tourism planning and management. This chapter is co-authored by Carla Fraga and Airton Rodrigues. The fourth chapter is authored by Pavlos Arvanitis and discusses air connectivity related with leisure tourism. The final chapter of this part is written by David Brown and Sharon Wilson, and explores the way Munro-baggers – hillwalkers – aim to climb in all the mountains of Scotland over 3000 ft, whilst it provides a hierarchy of themselves and others as serious leisure participants.

1 An exploration into Gen Ys' attitudes and behaviour towards volunteering whilst backpacking

Elena Jelaca, Iride Azara, and Eleni Michopoulou

Abstract

This study focuses on Generation Ys' attitudes and behaviour towards engaging in volunteer tourism whilst backpacking. To that end, we first examine Gen Ys' generational characteristics and the predominant attitudes and behaviours displayed by this generational cohort. Then the focus is shifted to understanding Generation Y as backpackers and their internal and external motivations. These motivations are queried under the prism of volunteer tourism; being seen as factors determining the level of engagement with volunteer tourism and overall backpacking behaviour while travelling. This chapter provides insights into the themes described above by examining the relevant tourism literature. Finally, it summarises the theoretical gaps in the extant literature and sets objectives for future research, whilst signposting authors to key literature sources.

Keywords:

Generation Y, millennials, backpacking, volunteering, tourism, motivation

Introduction

Generation Y has been the last labelled and thoroughly explored generation (Benckendorff & Moscardo, 2010), and leisure and tourism appear to be of particular significance to them (Halliday & Astafyeva, 2014). Jennings et al. (2010) provided insights on backpackers within Generation Y, highlighting their nonchalant desire for relaxation and avid appetite in terms of new experiences. However, travel and tourism behaviour of Generation Ys has still not been adequately explored and is therefore faced with many contradictory claims (Benckendorf & Moscardo, 2010; Tulgan & Martin, 2001). Hence, there is a need for further research within this field in order to consolidate present knowledge and provide new insights into the behaviours and attitudes of Generation Y backpackers (Merriam & Tisdell, 2016; Ong & du Cros, 2012). The aim of this study is to explore Generation Ys' attitudes and behaviour towards engaging in volunteer tourism whilst backpacking by setting and answering the following objectives:

1. Establish generation Ys' characteristics as backpackers and their motivations and attitudes towards volunteering,
2. Identify their motivators and attitudes for their involvement in volunteering whilst travelling and
3. Evaluate the extent to which motivators and attitudes towards volunteering whilst travelling affect Generation Ys' backpacking behaviour. This study contributes to theory by identifying gaps in the extant literature and providing recommendations for future research.

Theoretical background

Generational theory originates from the United States of America but is widely relevant for anglophones (i.e. English speakers), whether native or non-native, with a cultural background related to the English language (Fields et al., 2008). Generations or generational cohorts are usually defined as: *'proposed groups of individuals who are born during the same time period and who experienced similar external events during their formative or coming-of-age years (i.e. late adolescent and early adulthood years).'* (Noble & Schewe, 2003: 979). Researchers whose field covers population influences on society use the term '*generation*' as a common denominator for people born within the same general timeframe and experience the same social or historical key events (Gursoy et al., 2008). The term has also been used to define behaviours and attitudes of a group of people different from the generations before or after it (Beirne, 2008).

What makes Generation Y so interesting, according to Salt (2006), is the potential to be a positive and hopeful generation, more community-minded, focused on modesty, achievement, teamwork, and good conduct (Beirne, 2008) unlike the two previous generations, which have usually been defined by negative trends. As a group, it is different from any other in history, wealthier, more numerous, better educated, more ethnically diverse and interested in positive social skills. Several key defining events, social conditions and distinguishing features have been suggested as responsible for the Y generation's characteristics (Corvi et al; 2007; Mitchell, 2005; Eisner, 2005; Donnison, 2007; Howe & Strauss, 2003; Lyons et al., 2005; Jorgenson, 2003; McCrindle, 2009; Clemmons, 2008; Walker et al., 2006; Krayewski, 2009; Saltzman, 2007).

Key influences

Canavan (2011) groups shaping influences into three key themes: *technology*, *globalisation* and *parenting*.

In terms of technology, the last 30 years has been a period of unique alteration from an industrial to a digital economy and culture, followed by the appearance of the 'digital native'. The Information Age has caused one of the most influential shifts in our society to date (Pendergast, 2007). Generation Y was the first generation born into the Information Age, consequently known as digital natives (Prensky, 2006). The rise of the internet and the emergence of globalisation has affected global popular culture, made the world easier to travel and introduced digital media. As a result of those events Generation Y is technology savvy, seeking constant change, flexible, interested in creativity and innovation, multi-cultural and tolerant of diversity.

With regards to globalisation, it had significant impact on the overall Generation Y personality, making them so far, the best educated and most culturally diverse generation, open-minded and flexible towards diversity and change (Pew Research Centre, 2010; Wolburg & Pokrywczynski, 2001) thus transcending geography and ethnicity (Donnison, 2007;

Pendergast, 2010). Being born into the 'age of terrorism', exposure to terrorist events and feelings such as fear, shock and amazement in their formative years, has led Generation Y to be especially concerned when it comes to safety and security issues. Random and indiscriminate victims of terror events, and their viral distributionof reports through the internet '*have resulted in a strong safety net being thrown around the Y Generation.*' (Pendergast, 2010). There is an emphasis on instancy, risk, safety and short-term satisfaction triggered by the global financial crisis of 2008 that hit members of the generation as they entered the workforce (Pendergast, 2010). More than one third of the complete cohort are from minorities thus creating a global mix and match culture with growing tolerance for diversity (Morton, 2002).

Considering parenting, Howe and Strauss (2009) state that it was permissive parenting that defined this cohort. Because of it, Generation Y members are described as having high levels of self-respect, are confident and optimistic (Pendergast, 2010) with strong belief in their uniqueness (Howe & Strauss, 2009). Many of them have grown up in a 'non-traditional' family (Huang & Petrick, 2009), which resulted in their opinions on family being unconventional. Being raised in times of economic growth and prosperity made them ambitious and in constant seek of status and prestige. Active Baby Boomer parenting with more divorced parent cases and more working mothers with scheduled lives and fewer siblings, has produced numerous Gen Y member features like confident, respectful of authority, close to parents and family, optimistic but in the same time sheltered, indulged, entitled, seeking constant feedback, and questioning authority (Mitchell, 2005). They are both self-reliant and team oriented (Sullivan & Heitmeyer, 2008). Gen Y members are financially dependent on parents and for that reason have been nicknamed as 'helicopter kids' (Salt, 2006). They value work-life balance but are willing to work hard at the expense of life (O'Reilly, 2006). They are education oriented, pressured, competitive but group focused as an outcome of higher levels of education which demanded staying in school or at home longer and marrying later (Corvi et al. 2007). Civic-minded, concerned about the world, entrepreneurial, willing to volunteer and being altruistic represent Gen Y traits attributable to exposure to terrorism, increased violence and rise of environmentalism (Pendergast, 2010).

Contradictions and confusions

Still, the scope of the research on Generation Y is constrained in many ways and thus inherent with contradictions and confusion about what is claimed. For instance, this cohort is perhaps the most demanding to scrutinize because of significant disagreement over the beginning and ending cohort years, the name of the cohort itself and its juvenility (Benckendorff & Moscardo, 2010). Donnison (2007) places beginning the years of Generation Y in a period between 1977 and 1983, whilst ending the years period is considered from 1983 to 2009. The complexity of defining dates creates difficulties. The size of the generation also varies widely, and so do its labels. Estimates in size for US cohorts vary between 76 million (Eisner, 2005) and 80 million (Mitchell, 2005) and they can make Generation Y smaller (McCrindle, 2009) or larger than other generations (Mitchell, 2005). As far as labels are concerned, there is a debate over Generation Y as a universally accepted label. Generation Y members are also known as Echo Baby Boomers, Net Generation and Millennials (Eisner, 2005), but from the cohort themselves, the most commonly used label is Generation Y, shortened to Gen Y (McCrindle, 2009).

There is also confusion about what is considered a defining social condition versus what is considered a characteristic resulting from a social condition. Example is given with the evidence that shows how Gen Y members stay longer at home because of education and thus delaying marriage (Corvi et al., 2007). Given the fact that a number of members from this cohort may have not yet experienced their defining moments (independence from the family, career choices, marriage) and are still in the process of formation of their worldviews and values, it is not uncommon nor unacceptable to find contradiction and confusion in the research results to date (Sugarman, 2001; Benckendorff & Moscardo, 2010).

Further contradictions are apparent when accounting for Baby Boomer's influence. From one point of view, divorced and working mothers are perceived as contributors to the greater autonomy and independence of their children, forcing them to look after themselves (ABS, 2006), whilst from another point of view, anxious parents created sheltered and spoiled members of cohort (Corvi et al., 2007). Also, while Eisner (2005), Howe and Strauss (2003), Corvi et al. (2007) and Lyons et al. (2005) claim that the most important motivation for Gen Y members is finding a better work-life balance, Jorgenson (2003) and NAS (2006) deny it, claiming that they are ambitious perfectionists willing to sacrifice work-life balance for career promotion. On the one hand, deeply individualistic when it comes to their values (Mitchell, 2005) but on the other, very sociable and group oriented (Clemmons, 2008; Lyons et al., 2005). The complex picture of Gen Y members is complimented by yet another confrontation in claims. By some they are described as respectful of authority (Corvi et al., 2007; Howe & Strauss, 2003; Eisner, 2005) and by others as questioning the authority (NAS, 2006; Mitchell,2005).

Although Generation Y is not yet fully understood, *'it is worth remembering that the concept of generational cohorts has been used to explain a number of phenomena, and there is no evidence to suggest that generational theory should be abandoned for this group'* (Benckendorff & Moscardo, 2010).

Generation Y as backpackers

Pearce (1990) characterized a 'backpacker' as a type of traveller that prefers budget accommodation, travels longer time and has organised his/hers travel schedule in a flexible and independent way with an accent on meeting other travellers and spontaneous activities. Even though the criteria are fairly vague, they did not constitute an obstacle for many backpacking studies deriving the definition from this source specifically. Australia was one of the first countries that recognized value in youth tourism, as early as 1995. However, a more established version of the 'backpacker' term can be found in Paris' (2010) work:

> *'A backpacker is someone who seeks out experiences associated with budget travel, puts an emphasis on social interaction with local people and fellow travellers, is independently organised and flexible within constraints of time, money, and destination, travels as long as possible within the constraints of their own life, and puts an emphasis on informal and participatory holiday activities.'*

Motivations

Motivations are intrinsic and extrinsic forces that encourage tourists to move from their original habitats to different places in order to satisfy their needs (Dayour, 2013). Fundamental research on backpackers' motivation dates back to the last century emphasizing the prevailing theme as an 'escape' (Pearce, 1990). Overall motivations were described as many-sided and diverse (Pearce,1993) including escaping both interpersonal and personal worlds, monotony of everyday life or breaking away from job, relationship and career responsibility (Jarvis,1994). Likewise, stress in everyday life has been acknowledged as a push factor for those wishing to experience freedom, adventure and independence (Elsrud, 1998; Cohen, 2003). Because of the expectations of society upon their return, backpacking experience is for many travellers seen as unique and unrepeatable opportunity (Jarvis 1994). Research on youth travel conducted by WYSE Travel Confederation (2014) highlights two main motivators for young travellers: interaction with local people, and experiencing everyday life in foreign country. Intention to experience the 'local' plays an important role in bringing together young travellers and local communities, thus having a direct economic, social and cultural effect on them. An example of this derives from Amazon region of Brazil where backpacker tourism generated sociocultural benefits, improved local employment opportunities and conservation of the location (Rodrigues & Prideaux, 2012).

Cavagnaro, Staffieri and Postma (2018) carried out research among Dutch young population ratifying the fact that Generation Y members are not a homogenous group despite all the characteristics they share, differentiating them from previous cohorts. In their research they connected value orientations to the meaning of travel for Generation Y members and discovered contrasting needs for travel. While some Generation Y members match the description of young travellers seeking only simple amusement, there is a significant group of young people strongly motivated by self-transcending values, and will therefore result in distancing themselves for relaxation and escapism as their primary needs for travel. This shifting trend towards purposeful travel is evident and has become more important in the recent years. Key evidence can be found in WYSE Work and Volunteer Abroad Study (2015) which indicated that 80% of young volunteer travellers felt positive outcomes for their host families, host organisations or friends in a host country. This trend has been noticed by Pearce et al. (2009) on a sample of 100 backpackers' opinions about future important trends. Among seven trends, 'Volunteering skills and services to an indigenous or local community' and 'Volunteering skills and services to an environmental cause or study' have been perceived as of high and moderate importance.

UNWTO and WYSE Travel Confederation in their report 'Power of youth travel' from 2016 have portrayed youngsters as more flexible when compared to other generations, and therefore more interested in visiting politically and environmentally stressed destinations. Their main travel motivators were interaction with local people and the experience of everyday life in a foreign country. They were keen towards experiencing 'the local' thus having direct cultural, social and economic impact on communities they visit. According to WYSE (2016), young travellers spend their money within the local communities thus, nearly 60% of youth travel budgets are spent in the destination.

Self-development and self-actualisation

More energetic and purposeful forms of travel like volunteering and language learning have emerged; which are focusing on self-development with an impact on the destination as a result. It has been indicated that Generation Y members seek specific tourism experiences which allow them to emphasize their complicity (Gardiner et al., 2014) and strive for authentic tailor-made experiences imbued with choice, freedom and flexibility (Huntley, 2006).

Furthermore, they are willing to skip on travel and accommodation costs in order to spend more at the destination (Cavagnaro ET AL., 2018), especially on value-providing signature experiences (Loker-Murphy & Pearce, 1995). It is a group which seeks leisure experiences for the self-actualisation, where personal realization comes first, pushed by the high demands and expectations of themselves (Fountain & Charters, 2010; Valentine & Powers, 2013; Debevec et al., 2013). Therefore, leisure and tourism serve as an important tool for obtaining personal goals (Halliday & Astafyeva, 2014). Their whole living philosophy has been described as hedonistic (Altinbasak-Farina & Guleryuz-Turkel, 2015), they live for the moment and are always looking for the next big thing when it comes to new experiences (Smith, 2016).

Volunteer tourism in the backpacking experience

Volunteer tourism, comprising elements of leisure, travel and volunteer work (Brown, 2005), is generally considered as paid experiences where tourists allocate their money in order to participate in arranged projects of special interest, with a prevailing focus on contribution to local environments (Broad, 2003). It can also be described as a form of ecotourism with an emphasis on altruistic travel motivators, more ideologically divergent and sustainable in comparison to traditional forms of tourism (Wearing, cited in Ooi & Lang, 2010). This way of travelling, where the travellers' purpose is to make a difference, is argued to be a most advanced and sophisticated aspect of alternative tourism (Brown & Morrison, 2003).

The benefits of volunteer tourism

Outcomes of volunteer tourism focus on contributions to host communities as they impose inevitable economic, socio-cultural, and environmental benefits for sustainable tourism. However, volunteer tourism presents mutual benefits for both community and participants; it has been shown to not only provide improvements on host communities in terms of environmental protection, but also offer social and personal benefits for participants (McGehee, 2002). Even though volunteer tourism is regarded as a potential environmental and social tool of change, several criticisms of organised volunteer tourism programmes have been made (Ooi & Lang, 2010). As noted by Raymond and Hall (2008), such organisations may unwittingly emphasize lack of power equality, whilst programmes may promote homogeneity of host community through simplified a portrayal and minimal interaction with the Third World (Simpson, 2004). Despite the criticism, volunteer tourism demonstrates the capability for enhancement of tourism sustainability because of its increasing recognition as a way of accountable tourism approach (Raymond & Hall, 2008).

The ideological explanation defines the volunteer tourist as an individual who, because of different reasons, undertakes holidays in an arranged way, that might include mitigating material poverty of some society groups, environmental restoration or exploring aspects of the environment or society (Wearing, 2001). Motivations for engagement in volunteer tourism, appear to be numerous and different, such as with backpacking (Ooi & Lang, 2010). However, the key repeating theme within volunteer tourism literature concentrates on a holiday with a difference (Coghlan, 2006) and giving back to the community (Brown, 2005). Albeit at first altruistic, these motivations may be a tool for boosting tourists' own image (Gray & Campbell, 2007; Wearing, 2001). Labelled as altruistic individualism (Beck & Beck-Gernsheim, 2002), this cohesion of altruism and ego-enhancement highlights both motivating factors as equally important when engaging in volunteer tourism.

Another important and well documented key motivating factor for engagement in volunteering tourism and backpacking lies in its ability to enable personal growth and self-awareness (Ooi & Lang,2010).

Through interaction and challenges they are faced with, individuals have the chance to grow and explore themselves as a result of their altruistic efforts (Matthews, 2008; Wearing, 2001). The effects of aid given to others manifest as an alteration of individuals' life perception and worldview (McGehee & Santos, 2005) with individuals being perceived as more open-minded and less self-centered. The literature also highlights further development of skills and experience as an outcome of participation in volunteer tourism in order to enhance the curriculum vitae and to help oneself (Galley & Clifton, 2004). Various studies have found the core essence of volunteering, such as working with contrasting nationalities (Wearing, 2001), interaction with people of the same values and interests (Raymond and Hall, 2008) and experiencing local culture (Broad, 2003), to be instrumental to the volunteering experiences from the tourism perspective. Connection to nature also appears to be a highly influential motivator; immersed in the natural environment volunteer tourists strive to preserve and improve it. They promptly learn about it from the locals (Wearing & Neil, 2000; Brown, 2005), thus assisting in the production of wanted genuine experiences (Brown, 2005).

In examining different motivators and outcomes of volunteer tourism, we cannot but acknowledge its inherent potential for transformation. Mezirow's transformational learning theory has been applied to various disciplines, and it is important to consider also here, as there is evidence of a link between tourism and personal transformation (Mezirow, 2009; Laing & Frost, 2017). Travel and tourism are identified as drivers of personal transformation, offering an opportunity for self-discovery (Laing & Frost, 2017), accompanied with cathartic and long-term lasting life changing impacts (Zahra & Mcintosh, 2007). The transformative power of tourism lies in the upcoming challenges travellers are confronted with, in a hitherto unknown environment. Challenge can be described as a culture shock recognized in sojourn studies (Lyon, 2002), culture confusion found in studies of backpackers in Sri Lanka and India (Hottola, 2004) or disorienting dilemma from transformational learning theory (Mezirow, 1978). A common denominator exists in the ability of a traveller, confronted with a situation in which existing values and learned behavioural responses do not function, to overcome confusion and disorientation by the process of adaptation, usually accompanied by behavioural, psychological and convictional transformations (Nowaczek, 2013). To trigger the transformation, travellers should have a desire to transform (Ross,

2010); or not necessarily, as confrontation with disorienting dilemmas cannot be anticipated or planned (Reisinger, 2013).

New environments also play a key role in travellers' transformation as they can provide significant contrast from the initial one, thus stimulating contemplation. Reisinger (2013) identified several surroundings suitable for transformation purposes such as nature, adventure, farming, culture, wellness, and education. For example, nature serves as a reminder of the human littleness; adventure when combined with fear of death results in spiritual growth; farming as an indicator of pure life; culture as an immensity of human talent; wellness as a tool of reconnecting with oneself; and education as an assistant for intercultural understanding. Hence, certain types of tourism experiences such as religious tourism, educational tourism, eco- and nature-based tourism, volunteer tourism and developmental, study abroad and backpacking tourism can serve as settings for augmenting travellers' transformative potential (Lean, 2009).

Conclusions

The majority of the research conducted on Generation Y concentrates more on purchasing behaviour, attitudinal and demographic characteristics rather than its behaviour while travelling (Huang & Petrick, 2009), distinguishing it as being relatively underdeveloped (Richards & Wilson, 2004), especially from the psychological and environmental standpoints (Richards & King, 2003; Hopkins, 2016). Characteristics of this unique cohort have been extensively researched, but very little is known of their travel and tourism behaviour, and many contradictory claims have been made (Benckendorff & Moscardo, 2010). This study contributes to the literature by bringing together and examining existing knowledge that sits in the crossroads between volunteer tourism, backpacking tourism, and Gen Y. The theoretical contribution therefore lies in the establishment of explicit links between these three areas and highlighting knowledge gaps.

Hence, there is a need for further research in this field to consolidate the present knowledge and provide new insights following the rare studies that contribute to our understanding of Gen Y backpackers (Ong & du Cros, 2012). Table 1.1, provides recommendations for further research by offering a summary of identified gaps in the extant literature, accompanied with relevant objectives and some indicative key sources.

Table 1.1: Theoretical gaps, objectives and indicative key literature

Theoretical gap	**Objective**	**Indicative literature**
Generational theory and Generation Y's characteristics		
Gap: Very little is known about their behaviour while travelling; Benckendorf & Moscardo (2010)	Establish Generation Ys' characteristics as backpackers and their motivations and attitudes towards volunteering	Pendergast (2010), CRC for Sustainable Tourism Pty Ltd, (2009) Paris, (2010)
Generation Y as backpackers		
Gap: Need for further research in order to consolidate present knowledge and new insights to contribute the breadth of generation Y among backpackers; Ong & du Cros, (2012)	Analyse characteristic behaviour of Generation Y as backpackers and identify their motivators and attitudes for their involvement in volunteering whilst travelling	WYSE Travel Confederation (2014) Cohen (2003); Cohen (2004) Staffieri & Cavagnaro, (2018) Salvaggio (2016); Majstorovic et al., (2013) Cavagnaro, Staffieri & Postma, (2018) McGehee & Santos, (2005) Ooi & Laing, (2010); Hottola, (2004)
Volunteer tourism as part of backpacking experience		
Gap: Why younger generations, seek alternative volunteer tourism experiences; Zahra & McIntosh, (2007)	Evaluate the extent to which motivators and attitudes towards volunteering whilst travelling affect Generation Ys' backpacking behaviour	**Personal transformation** WYSE Work and Volunteer Abroad Study (2015) Matthews, (2008) Coghlan (2006); Brown (2005) Galley & Clifton, (2004) Ooi & Laing, (2010) Gray & Campbell, (2007); Wearing, (2001) **Sustainable travel** CRC for Sustainable Tourism Pty Ltd, (2009) Reisinger, (2013) Devlin, (1998); Deville, (2011) Hopkins, (2016) **Future employability** Cohen, (2003) Steimel, (2017) Litauen & Miller, (2012) CRC for Sustainable Tourism Pty Ltd, (2009)

Besides the theoretical implications, this study also bears implications for practitioners. As Ooi and Laing (2010) indicate, there is an overlap of backpacker and volunteer tourism motivations and understanding this consumer market can yield synergies for designing leisure and tourism activities. The alignment of these two niche tourism segments may facilitate the promotion of volunteer platforms and encourage altruistic travel experiences among more diverse tourism audiences, with wider benefits for the tourism industry as a whole. Volunteer tourism organisations could also see backpackers as a potential new market and target them through specific tailor-made opportunities based on their motivations and attitudes towards volunteering.

In conclusion, it is important to understand why the younger Generations X and Y are seeking alternative volunteer tourism experiences and how this will shape different forms of consumption – and tourism consumption in particular. This understanding will provide us with a new perspective of backpacker volunteer tourists, which will help address the challenges in the future direction of the tourism sector; by demystifying a market segment that manifests as both 'consumer' and 'employee' and develop strategies to better attract, interact and co-create experiences that deliver value to all involved.

Highlights

- There is contradictory evidence in terms of the Gen Y predominant characteristics. For instance, whilst some studies claim that it is important for Gen Y members to find better work-life balance, others deny it by claiming that they are ambitious perfectionists willing to sacrifice work-life balance for career promotion.
- Backpacking is a popular activity amongst Gen Y, but the motivations vary.
- Volunteer tourism presents mutual benefits for both community and participants and is an activity that is often undertaken whilst backpacking; also highlighting commonalities in motivations to engage in either/both of these types of tourism.

Questions

1 Considering Ooi and Laing's (2010) work on the overlap of backpacker and volunteer tourism motivations, where alignment of these two niche tourisms may facilitate the promotion of volunteer platform, encouraging altruistic travel experiences among a more diverse tourism audience, try to think of beneficial outcomes for the tourism industry in general.

2 Consider the implications for industry practice regarding backpackers who are willing to undertake volunteering opportunities whilst backpacking. How can they be effectively targeted?

3 Will backpacking and volunteer tourism still be in high demand from Generation Y members after Covid-19 crisis? Will there be a shift in Generation Y members' values towards priorities in their lives? Will Gen Y altruistic traits (helping others, helping in highly affected areas) overrun their self-centred traits? Which Generation Y characteristic will prevail – adventurous or cautious? Explain and substantiate with literature.

References

ABS (2006). *A Picture of the Nation – 2070.0 2006*. Australian Bureau of Statistics, Canberra.

Altinbasak-Farina, I. & Guleryuz-Turkel, G. (2015). Identifying the needs of Gen Y by exploring their value systems: a qualitative study. *International Journal of Trade, Economics and Finance*, 6(6), 290-296.

Beck, U., & Beck-Gernsheim, E. (2002). *Individualization. Institutionalized Individualism and its Social and Political Consequences*. London: Sage.

Beckendorff, P., & Moscardo, G. (2010). Understanding Generation Y tourists: Managing the risk and change associated with a new emerging market. *Tourism and Generation Y*. Cambridge, MA: CAB International.

Beirne, M. (2008). Generation Gap. *Brandweek* 49, 16-20.

Broad, S. (2003). Living the Thai Life—A case study of volunteer tourism at the Gibbon Rehabilitation Project, Thailand. *Tourism Recreation Research*, 28(3), 63-72.

Brown, S. (2005). Travelling with a ourpose: Understanding the motives and benefits of volunteer vacationers. *Current Issues in Tourism*, 8(6), 479-496.

Brown, S. & Morrison, A. (2003). Expanding volunteer vacation participation an exploratory study on the mini-mission concept. *Tourism Recreation Research*, 28(3), 73-82.

Canavan, B. (2011). Displacement of youth from the Isle of Man: the role of house price inflation. *Island Studies Journal*, 6(2), 203-226.

Cavagnaro, E., Staffieri, S. & Postma, A. (2018). Understanding millennials' tourism experience: values and meaning to travel as a key for identifying target clusters for youth (sustainable) tourism. *Journal of Tourism Futures*, 4(1), 31-42.

Clemmons, D. (2008). VolunTourism and the M3 travelers.

Coghlan, A. (2006). Volunteer tourism as an emerging trend or an expansion of ecotourism? A look at potential clients' perception of volunteer tourism organisations. *International Journal Nonprofit Volunteer Sector Mark*, 11, 225-237.

Cohen, E. (2003). Backpacking: Diversity and change. *Journal of Tourism and Cultural Change*, 1(2), 95-110.

Cohen, E. (2004). *Contemporary Tourism: Diversity and Change*. Amsterdam. Elsevier.

Corvi, E., Bigi, A. & Ng, G. (2007). The European Millennials versus the US Millennials: similarities and differences. *Working paper.*

CRC for Sustainable Tourism Pty Ltd (2009). *Evolution of the backpacker market and the potential for Australian tourism*. Gold Coast, Queensland. https://www.researchgate.net/publication/228670523_Evolution_of_the_Backpacker_Market_and_the_Potential_for_Australian_Tourism (Accessed 15 Dec. 2019).

Dayour, F. (2013). Are backpackers a homogeneous group? A study of backpackers' motivations in the Cape Coast-Elmina conurbation, Ghana. *European Journal of Tourism, Hospitality and Recreation*, 4(3), 69-94.

Debevec, K., Schewe, C., Madden, T. & Diamond, W. (2013). Are today's Millennials splintering into a new generational cohort? Maybe! . *Journal of Consumer Behaviour*, 12(1), 20-31.

Deville A. (2011). *Alice in WWOOFerLand: Exploring symbiotic worlds beyond tourism*. (Doctoral dissertation). University of Technology, Sydney, Australia.

Devlin, C. (1998). WWOOFing for your Supper: the WWOOF experience. *Permaculture International Journal*. 29-31.

Donnison, S. (2007). Unpacking the millennial's: a cautionary tale for teacher education. *Australian Journal of Teacher Education* 32(3), 1-13.

Eisner, S.P. (2005). Managing Generation Y. S.A.M *Advanced Management Journal*, 70, 4-15.

Elsrud, T. (1998). Time creation in traveling: The taking and making of time among women backpackers. *Time and Society*, *7*, 309–334.

Fields, B., Wilder, S., Bunch, J. & Newbold, R. (2008). *Millenial Leaders: Success Stories from Today's Most Brilliant Generation Y Leaders*. Ingram Publishing.

Fountain J., & Charters, S. (2010). *Younger wine tourists: A study of generational differences in the cellar door*. Paper presented at the International Wine and Tourism Conference, Margaret River, Australia.

Galley, G. & Clifton, J. (2004). The motivational and demographic characteristics of research ecotourists: Operation Wallacea volunteers in Southeast Sulawesi, Indonesia. *Journal of Ecotourism*, 3(1), 69-82.

Gardiner, S., Grace, D. & King, C. (2014). The generation effect. *Journal of Travel Research*, 53(6), 705-720.

Gray, N. & Campbell, L. (2007). A decommodified experience? exploring aesthetic, economic and ethical values for volunteer ecotourism in Costa Rica. *Journal of Sustainable Tourism*, 15(5), 463-482.

Gursoy, D., Maier, T. & Chi, C. (2008). Generational differences: An examination of work values and generational gaps in the hospitality workforce. *International Journal of Hospitality Management*, 27(3), 448-458.

Halliday, S. & Astafyeva, A. (2014). Millennial cultural consumers: co-creating value through brand communities. *Arts Marketing: An International Journal*, 4(1/2), 119-135.

Hopkins, D. (2016). Can environmental awareness explain declining preference for car-based mobility amongst generation Y? An examination of learn to drive behaviours. *Transportation Research Part A: Policy & Practice.*

Hottola, P. (2004). Culture Confusion: Intercultural Adaptation in Tourism. *Annals of Tourism Research*, 31(2), 447–466. https://doi.org/10.1016/j.annals.2004.01.003

Howe, N., & Strauss, W. (2003). *Millennials Go to College.* Washington, DC: American Association of Collegiate Registrars and Admissions Officers.

Howe, N., & Strauss, W. (2009). *Millennials Rising: The next great generation.* New York: Vintage Books.

Huang, Y.C., & Petrick, J.F. (2009). Generation Y's travel behaviours: A comparison with baby boomers and Generation X. *Tourism and Generation Y.* Cambridge, MA: CAB International.

Huntley, R. (2006). *The World According to Y: Inside the New Adult Generation.* Crows Nest: Allen & Unwin.

Jarvis, P. (1994). Coping in adolescence: Empirical evidence for a theoretically based approach to assessing coping. *Journal of Youth and Adolescence*, 23(3), 359-371.

Jennings, G., Cater, C., Lee, Y-S., Ollenburg, C., Ayiling, A. & Lunny, B. (2010). Generation Y: Perspectives in youth adventure travel experiences in an Australian backpacker context. In P. Benkendorff, G. Moscardo, & D. Pendergast (Eds.) *Tourism and Generation Y*, pp. 58-72. Oxfordshire: CABI.

Jorgenson, B. (2003). Baby Boomers, Generation X and Generation Y? Policy implications for defence forces in the modern era. *Foresight* 5, 41-49.

Krayewski, K. (2009). Generation Y and why they matter. http://internationalaffairs.suite101.com/article.cfm/generation_y

Laing, J. H., & Frost, W. (2017). Journeys of well-being: Women's travel narratives of transformation and self-discovery in Italy. *Tourism Management*, 62, 110-119.

Lean, G. L. (2009). Transformative travel: Inspiring sustainability. In Bushel, R. & Sheldon, P. (Eds.) *Wellness and tourism: Mind, body, spirit, place*. New York: Cognizant.

Litauen, B.C., & Miller J. (2012). *Corporate Social Responsibility and Generation Y*. pp 43-44.

Loker-Murphy, L. & Pearce, P. L. (1995). Young budget travelers: Backpackers in Australia. *Annals of Tourism Research*, 22, 819–843.

Lyon, D. (2002). Everyday surveillance: Personal data and social classifications. *Information, Communication & Society*, 5(2), 242-257.

Lyons, S., Duxbury, L. & Higgins, C. (2005). An empirical assessment of generational difference in work-related values. *Human Resources Management* 26, 62-71.

Majstorovic, V., Stankov, U. & Stojanov, S. (2013). The presence of backpacking tourism in Europe. *Turizam*, 17(4), pp.145-154.

Matthews, A. (2008). Negotiated selves: Exploring the impact of local-global interactions on young volunteer travellers. *Journeys of Discovery in Volunteer Tourism*. Wallingford: CABI Publishing.

McCrindle, M. (2009). Seriously cool: marketing, communicating and engaging with the diverse generations. http://www.mccrindle.com.au

McGehee, N.G. (2002). Alternative tourism and social movements. *Annals of Tourism Research*, 29, 124-143.

McGehee, N.G., & Santos, C.A. (2005). Social change, discourse and volunteer tourism. *Annals of tourism research*,32, 760-779.

Merriam, S. B., & Tisdell, E. J. (2016). *Qualitative Research: A Guide to Design and Implementation*. 4th edn. San Francisco, CA: Jossey-Bass.

Mezirow, J. (1978). Perspective Transformation. *Adult Education Quarterly, 28*, 100-110. dx.doi.org/10.1177/074171367802800202

Mezirow, J. (2009). Transformative learning theory. In J. Mezirow, and E. W. Taylor (Eds), *Transformative Learning in Practise: Insights from Community, Workplace, and Higher Education*. Jossey-Bass.

Mitchell, S. (2005). The Millennial generation: its demographics, market preferences, productivity and global impact. In *Managerial Excellence and Networked Collaboration for Global Competitiveness*. International Training Center, Chula Vista, California.

Morton, L.P. (2002). Targeting Generation Y. *Public Relations Quarterly*, 47, 16-26.

NAS (2006). Generation Y: The Millennials. NAS recruitment. http://www.nasrecruitment.com/talenttips/NASinsights/Generation Y.pdf

Noble, S. M. & C. D. Schewe (2003). Cohort segmentation: an exploration of its validity. *Journal of Business Research*,56, 979-987.

Nowaczek, A., M., K. (2013). The next stage of eco-travel: facilitating transformation through personal ethics. In Reisinger, Y. (Ed.) *Transformative Tourism: Tourist perspectives*. Wallingford: CABI. pp. 169-179.

Ong, C.E., & du Cros, H. (2012). The post-Mao gazes: Chinese backpackers in Macau. *Annals of Tourism Research* 39(2), 735–754.

Ooi, N. & Laing, J. (2010). Backpacker tourism: sustainable and purposeful? Investigating the overlap between backpacker tourism and volunteer tourism motivations. *Journal of Sustainable Tourism*, 18(2), 191-206.

O'Reilly, C. C. (2006). From drifter to gap year tourist: Mainstreaming backpacker travel. *Annals of Tourism Research, 33*(4), 998–1017.

Paris, C. (2010). Backpacker motivations: A travel career approach. *Journal of Hospitality Marketing & Management*, 19(3), 244-259.

Pearce, P. L. (1990). *The Backpacker Phenomenon: Preliminary Answers to Basic Questions*. Townsville: James Cook University.

Pearce, P. L. (1993). Fundamentals of tourist motivation. In Pearce,, D.G. & Butler, R.W. (eds.) *Tourism Research: Critiques and Challenges*, pp. 85-105. London: Routledge and Kegan Paul.

Pearce, P.L., Murphy, L., & Brymer, E. (2009). *The Evolution of Backpacking and the Potential for Australian Tourism.* Brisbane: CRC Sustainable Tourism.

Pendergast, D. (2007). The MilGen and society. In: Bahr, N. & Pendergast, D. (eds) *The Millennial Adolescent*. Australian Council for Educational Research, Camberwell, Va, Australia, pp. 23-40.

Pendergast, D. (2010). Getting to Know the Y Generation. In: Benckendorff, P., Moscardo, G., & Pendergast, D. (eds) *Tourism and Generation Y*. CAB International, Cambridge, 85-97.

Pew Research Centre. (2010). *Millenials: Confident, connected, open to change.* https://www.pewresearch.org/wp-content/uploads/sites/3/2010/10/millennials-confident-connected-open-to-change.pdf (Accessed 17 December 2019).

Prensky, M. (2006). Listen to the natives. *Educational Leadership*, 63, 8-13.

Raymond, E.M., & Hall, C.M. (2008). The development of cross- cultural mis(understanding) through volunteer tourism. *Journal of sustainable tourism*, 16(5), 530-543.

Reisinger, Y. (2013). Transformation and transformational learning theory. In Reisinger, Y. (Ed.) *Transformative tourism: tourist perspectives*. Wallingford: CABI. pp. 17-26.

Richards, G. & King, B. (2003). Youth travel and backpacking. *Travel and Tourism Analyst*, 6, 1-23.

Richards, G., & Wilson, J. (2004). *The Global Nomad: Backpacker travel in theory and practice*. Clevedon: Channel View Publications.

Rodrigues, C. & Prideaux, B. (2012). Backpacker tourism in the Brazilian Amazon: challenges and opportunities. In G. Lohmann & D. Dredge (Eds.), *Tourism in Brazil: environment, management and segments* (pp. 123-140). London: Routledge.

Ross, S., L. (2010). Transformative travel – an enjoyable way to foster radical change. *ReVision.* 31(1), 54-61.

Salt, B. (2006). *The Big Picture: Life, work and relationships in the 21st century.* Prahran: Hardie Grant Books.

Saltzman, D. (2007). 'Youthanize' your travel agency. *Travel Trade,* 9 November. http://www.traveltrade.com/news_article.htm?id=238

Salvaggio, M. J. (2016). Bursting the backpacker bubble: exploring backpacking ideology, practices, and contradictions. *UNLV Theses, Dissertations, Professional Papers, and Capstones.* Available at: https://digitalscholarship.unlv.edu/thesesdissertations/2900

Simpson, K., (2004). 'Doing development': the gap year, volunteer-tourists and a popular practice of development. *Journal of International Development,* 16(5), 681-692.

Smith K. M. (2016). *Issues in Cultural Tourism Studies.* 3rd edn. New York: Routledge.

Staffieri S., & Cavagnaro E. (2018). Youth travel experience: An analysis of the relations between motivations, satisfaction and perceived change. *WIT Transactions on Ecology and the Environment, 227,* 95-106.

Steimel, S. (2017). Skills-based volunteering as both work and not work: A tension-centered examination of constructions of 'volunteer'. *Voluntas: International Journal of Voluntary and Nonprofit Organizations,* 29(1), 133-143.

Sugarman, L. (2001). *Life-span Development.* 2nd edn. London: Routledge.

Sullivan, P. & Heitmeyer, J. (2008). Looking at Gen Y shopping preferences and intentions: exploring the role of experience and apparel involvement. *International Journal of Consumer Studies, 32,* 285-295.

Tulgan, B., & Martin, C. A. (2001). *Managing Generation Y: Global citizens born the late seventies and early eighties.* Amherst, MA: HRD Press.

United Nations World Tourism Organisation and World Youth Student and Educational Travel Confederation (2016). *Affiliate members global reports, volume thirteen - the power of youth travel.* Madrid: UNWTO. https://www.wysetc.org/wp-content/uploads/2016/03/Global-Report_Power-of-Youth-Travel_2016.pdf (Accessed 16 Dec. 2019).

Valentine, D. B., & Powers, T. L. (2013). Generation Y values and lifestyle segments. *Journal of Consumer Marketing,* 30(7), 597–606.

Walker, J.T., Martin, T., White, J., Elliott, R., Norwood, A., Mangum, C. & Haynie, L. (2006). Generational (age) differences in nursing students' preferences for teaching methods. *Journal of Nursing Education* 45, 371-374.

Wearing, S., (2001). *Volunteer Tourism: Experiences that make a difference.* Wallingford: CABI Publishing.

Wearing, S. & Neil, J. (2000). Refiguring self and identity through volunteer tourism. *Society and Leisure,* 23, 389 – 419.

Wolburg, J. & Pokrywczynski, J. (2001). A psychographic analysis of Generation Y college students. *Journal of Advertising Research,* 41(5), 33-52.

WYSE (2016). Facts and Stats. https://www.wysetc.org/about-us/facts-and-stats/ (Accessed 3 March 2020).

WYSE Travel Confederation. (2014). *Millennial Traveller: An insight into the general travel behaviours and attitudes of millennial travellers.* https://www.wysetc.org/wp-content/uploads/2014/11/Millennial-Traveller-executive-summary.pdf (Accessed 28 November 2019).

WYSE Work and Volunteer Abroad. (2015). *Development through volunteering overseas.* https://www.wysetc.org/wp-content/uploads/sites/8/2015/05/Development-through-volunteering-overseas.pdf (Accessed: 28th January 2020).

Zahra, A. & Mcintosh, A. (2007). Volunteer tourism: Evidence of cathartic tourist experiences. *Tourism Recreation Research,* 32(1), 115-119.

2 Well-being experiences in nature: An emerging form of tourism

Alba Font-Barnet and Marta Gemma Nel-lo Andreu

Abstract

Tourism is a dynamic, constantly evolving activity that requires detecting consumer trends so that companies and destinations can adapt by offering new and improved competitive products and experiences. For the post-COVID-19 era, dominant projected trends suggest that consumers will be more attracted to authentic, environmentally responsible experiences in less touristy destinations and wide-open spaces. Natural areas and protected spaces rank among the clearest beneficiaries of such trends because they combine relaxation with environments conducive to social distancing as well as offering direct, positive impacts on human health and well-being. For all of those reasons, such spaces not only have the potential to offer new experiences closely related to well-being and the search for overall health but also face challenges to that end. Against that background, this chapter examines experiential tourism in relation to improved holistic well-being and provides examples of experiences in protected natural areas worldwide that offer physical, mental, social, environmental, spiritual, and/or emotional well-being.

Keywords

Experience, nature, protected area, tourism, well-being.

Introduction

Tourism is a dynamic, constantly evolving activity that requires detecting consumer trends so that companies and destinations can adapt by offering new and improved competitive products and experiences. In that light, consumer behavior draws the lines of action in the tourism sector, with special emphasis on the challenges that the sector faces during emergency situations. Amid the complexity of crises and situations of change in tourism, identifying and analyzing elements of a modulating, interrelated, interdependent nature can help to mark the evolution of the sector, as well as its origin and destination markets, in both the short and long term (Hernández & Padrón, 2020; UNWTO, 2021). This time around, the crisis is the COVID-19 pandemic, which has hit the tourism sector harder than most. As the pandemic continues to alter the behavior of tourists and patterns of mobility worldwide, its effects have also accelerated the reconstruction of altered forms of tourism. Thus, with change comes opportunity, as the adage goes—in tourism sector's case, the opportunity to rethink and redevelop tourism as an asset to the shift toward more resilient, sustainable communities (Euruni, 2020; One Planet Sustainable Tourism Programme, 2020).

Rogan (2020) has predicted that 2021 will be a year in which renewal and reconnection assume new value. For tourism, the forecast means that today's travelers will seek out unique, safe and eco-friendly experiences. The UNWTO's observations indicate a trend of proximity-based tourism, along with nature getaways and outdoor experiences. For that reason, its global guidelines to restart tourism include promoting and developing new, segmented products targeted at individual travelers and small groups with special interests in nature, sports, culture, and gastronomy (UNWTO, 2020). Added to that, Riley (2020) has projected that after the COVID-19 pandemic, people will be more appreciative of and attracted to meaningful experiences, feel greater responsibility for the environment, and, in turn, seek out adventure travel, which by its very nature has all of the makings of a post-pandemic antidote, given its focus on less touristy destinations and wide-open spaces.

TripAdvisor (2020) has affirmed that nature is one of the pandemic's clearest beneficiaries, one that offers relaxation in an environment conducive to social distancing. Beyond that, tourism based on and in connection with nature is directly related to well-being and holistic health for society. Experiences that enhance well-being can be used to transform not only the self but also one's attitudes and relationship with the environment (Pope, 2018). With such environmental awareness, societies can realize a more secure future for wildlife, higher-quality ecotourism, and greater visitor satisfaction (Newsome, 2020). Tourism and travel trends suggest that the market for trips focused on well-being has grown exponentially, while tourism in protected areas has grown progressively over the past 100 years, all within a sector that has steadily matured into one of the largest industries in the world (Azara et al., 2018). Worldwide, natural places and protected areas are recognized as destinations with all of the elements and challenging new experiences needed to promote the health and well-being of tourists, as well as local communities and the destinations themselves (Russell et al., 2013; Azara et al., 2018).

Against that background, the aim of this research is to exemplify and demonstrate the trends in well-being tourism in nature with good practices. This chapter first explores tourism experiences in relation to well-being and nature, a current trend, an interconnected relation, and an emerging form of tourism worldwide. Next, considering the physical, mental, social, environmental, spiritual, and emotional dimensions, suggested by Global Wellness Institute (GWI, 2018), it provides examples of experiences in natural areas that are seeking these dimensions. All this will allow a synthetic discussion that ties everything together, and that shows the interrelation between concepts, experiences, and well-being dimensions. Finally, the managerial implications and conclusion section summarize the value and results of this chapter. It ends with the three keypoints and the three questions addressed to researchers and students.

Experience, well-being, and nature: An interconnected relationship

Although the growing need to design and offer differentiated tourist experiences has increasingly been taken into account in the tourism sector (Brent-Ritchie et al. 2011; Sugathan & Ranjan, 2019), its origins date back to the mid-1970s (Csikszentmihalyi, 1975). The idea gained significant strength with the work of Pine and Gilmore (1999), who defined *experience* as occurring within a person who is engaged with an event on an emotional, physical, spiritual, or intellectual level. Along similar lines, O'Dell (2007) and Coelho et al. (2018) have referred to an *experience* is subjective, intangible, continuous, highly personal phenomenon. Extending the concept to tourism, Carballo et al. (2015) have proposed that tourists who travel to destinations do not make trips in a merely physical sense, for their true trips are internal and lie in how they perceive the destination, its people, and the impact of the subjective experience. With both of those ideas in mind, Selstad (2007) has defined the *tourist experience* as a combination of novelty and familiarity involving the individual pursuit of identity and self-realization.

To be sure, the tourist experience is a complex psychological process. Holistic and multiphasic, it encompasses an array of processes and dynamics that begin even before the trip and continue during the trip, in the place where the experience occurs, and when the trip, once complete, is continually (re-)evaluated to take stock of what was lived (Highmore, 2002; Almeida & Park, 2017). Creating the tourist experience, as such a complex phenomenon, requires considering multiple elements and intrinsic aspects before an effective, rewarding experience, with a wide range of opportunities at all levels, can be offered. Such an experience has to be unique, stimulate all of the senses, be emotional and immersive, and offer a story to be recounted (Thellier, 2011). In a sense, tourist experiences are only memorable when they are authentic and afford the opportunity to play an active role, forge social relationships, grow personally, and learn.

The interaction between the tourist, as the chief actor, and the destination, where the tourist experience occurs, is the foundation for constructing the experience (Vada et al., 2019; Same & Larimo, 2012). Although the tangible, physical elements of destinations are important, the key to the tourism experience is ensuring the emotional involvement of tourists, as well as satisfying and responding to different expectations and motivations, including escape, adventure, relaxation, education, training, personal fulfillment, health, and well-being (Li, 2000; Quinlan & Carmichael, 2010) (Figure 2.1).

According to GWI (2018), *well-being* refers to the active pursuit of activities, choices, and lifestyles that lead to a holistic state of health. In providing a combination of pleasure, hedonism, altruism, and lived experience, experiential tourism intersects with well-being in both its hedonic and eudemonic forms. Whereas hedonic well-being involves positive emotions, happiness, and pleasure during an activity, eudaimonic well-being focuses on personal growth, learning, and skills development, which confers positive effects that may persist well after the activity has ended (Pearce and Packer, 2013).

For some time now, researchers have demonstrated that natural environments can have direct, positive impacts on well-being, with stimulating effects for human health and general welfare (Bowler et al., 2010; Ryan et al., 2010; Cvikl, 2016). In that view, so-called "EcoWellness" was defined as a sense of appreciation, respect for, and aware of nature that results in feelings of connectedness with the natural environment and the enhancement of holistic wellness (Ryse & Meers, 2012). In this way, it is stated that there is an interaction between the well-being and experience from the holistic and integrative vision, and nature can become the link and place to develop experiences to achieve well-being.

Figure 2.1: The nature: linking experience and well-being.
Source: Authors' own elaboration

Well-being experiences in protected areas

Protected natural areas are delimited, safe places with rich ecosystems nurtured under a consolidated management structure and that provide ample space for recreation (Parks Victoria, 2017; Europarc, 2018); that's why these are key areas for the development of well-being tourist experiences. As such, it is imperative that protected areas are prepared to accommodate the rising interest in nature and to guide well-being tourism toward positive outcomes (One Planet Sustainable Tourism Programme, 2020).

In turn, the well-being afforded by nature has been emphasized in planning for protected areas under the worldwide Australia-based Healthy Parks, Healthy People program, which offers activities and strategies for improving human well-being and the conservation of parks at the same time. Because well-being encompasses a diverse set of activities and interests, including preventive health, beauty, fitness, personal growth, and nature (Smith & Puczó 2016; GWI, 2018), a wide range of activities for both mind and body can be developed in protected areas, whether for solo or group travelers, all in wonderful landscapes and surroundings that can offer spiritual as well as emotional benefits and connections with the environment and the self (Azara et al., 2018; Romagosa et al., 2015). In what follows, we provide a brief tour of well-being tourist experiences that combine the environmental, physical, mental, social, physical, spiritual, and/or emotional dimensions developed in and for natural protected areas.

Environmental dimension

Parks are perfect places to develop the environmental dimension, which emphasizes the value of local ecosystems and human involvement in the natural environment (Maller et al., 2002; Albuqerque et al., 2019). Beyond that, the dimension can afford mental and spiritual benefits (e.g., reduction of stress and personal growth), physical health (e.g., sports and fitness activities), and social opportunities, at least when in the company of others. As an example, the Wider Peak District Cycle Strategy is an interesting proposal developed in public–private collaboration for the UK's Peak District National Park (Peak District, 2020). According to its promoters, cycling is an ideal activity that provides the speed to explore the diversity of landscapes while remaining slow enough to enjoy the sights, smells, and sounds of the countryside. The aim of the program is to promote cycling due to the benefits that it brings, including education, enjoyment, athletic challenges, and tourism, not to mention sustainable transport, carbon reduction, and economic growth in the area, but above all well-being, social relations and mental health.

Physical and social dimension

Along the same lines, the website of the U.S. National Park Service (USNPS) now includes a section, titled 'Trails and Hiking', about the network of trails and their physical conditions, accessibility, and type (e.g., good for children or pets), as well as the benefits of being in outdoor spaces:

> *"The warm sunshine on your face, the sound of the wind rushing through the trees overhead, and the soft earthy feel of the trail under your boots. Not only are these experiences enjoyable to have, but they're good for you, too. Hiking is proven to have many health benefits, ranging from physical exercise you get when out on the trail, to emotional or mental relief that comes from being in nature"* (USNPS, 2020).

In fact, the USNPS has identified so-called 'green exercise clinics' as a Healthy Parks, Healthy People promising practice. In that sense, a *green exercise* is a physical activity or exercise in direct exposure to nature that positively affects psychological and physical well-being and health (Loureiro & Veloso, 2017). Among the possible activities, many are usually performed in gyms and closed spaces, including yoga and tai chi, and by 'going green', so to speak, reinforce their numerous benefits for the body and mind (Gladwell et al., 2013). In August 2020, the USNPS (2020) also launched the 'Fitness Challenge Campaign' designed to increase levels of physical activity in the short or long term, all to improve physical health, boost self-esteem, promote confidence, and strengthen will, especially in populations that seldom use parks, including minorities, low-income households, and people with disabilities. One set of challenges, for example, could include walking 5 miles in a park or completing activities along a trail, either individually or in a group, the latter as a means to socialize and create bonds in the community.

The National Parks and Wildlife Service of New South Wales, Australia, has put forward some interesting proposals, including high-intensity training workouts with beach sprints along long stretches of unspoiled coastline in Termeil Beach, trail running in Dharug National Park as a pre-workout for competitive races, and, reaping the sea's medicinal benefits as part of thalassotherapy, swimming at Dudley Beach in the Glenrock State Conservation Area (Gillezeau, 2020). As that final example shows, thalassotherapy can now be categorized in forms beyond treatments based on marine environments into authentic wellness-oriented experiences (Smith & Puczó, 2016).

Mental dimension

On the topic of medicinal and health benefits, mindfulness offers various advantages for both individuals and the sites of their experiences (Dutt & Ninov, 2012). Ericson et al. (2014) have proposed promoting mindfulness as a policy that contributes to both sustainability and greater well-being. Practices of mindfulness consist of taking time to be aware of personal feelings, one's surroundings, and what can be heard, smelled, touched, and/or seen. The practice can be realized by, for instance, spending time in a forest and engaging all of the senses, which can help to ease negative feelings and boost positive ones (WWF-UK, 2020).

Along those lines, *shinrin-yoku* ('forest bathing'), as an emerging practice the world over, has perhaps become the most representative trend. Li (2018) defined forest bathing as a traditional practice of taking energy from the forest to improve one's health and reduce stress. Japan Travel (2020) promotes forest bathing as an excellent way to appreciate the landscapes that once stirred such a deep, spiritual connection and to remind us that nature has always been regarded as powerful and healing. Japan's most enchanting forests are in Yakushima Island National Park or Yoshino-Kumano National Park, each with a network of ancient forested pilgrimage routes. Of course, forest bathing is an experience that has spread to and been implemented at other national parks around the world, including natural areas in British Columbia, Canada (Ali, 2018); in the UK, which offer a beginner's guide to forest bathing in nature parks across the island (National Trust, 2020); in Victoria, Australia, where the government has developed a map platform with the different experiences for forest bathing on offer (Visit Victoria, 2020); and in Catalonia, Spain, where there is an important offer of forest baths certified and guided by Selvans Cooperative in natural areas (Selvans, 2020). For example, the Els

Ports Natural Park (PNP), located in the south of Catalonia, where these experiences have been carried out by Ports Experience, among many other well-being experiences (see the case study, p. 35). Because forest bathing began as a medical prescription in Japan, the USNPS (2020) has extended that philosophy into its 'Doctor-Led Activities', which take patients outdoors to harness the healing power of nature and get inspired to make positive changes using various tools and techniques: Park Prescriptions (ParkRx), Walk with a Doctor, Nature & Forest Therapy and Art Therapy.

Emotional and spiritual dimension

For those in search of experiences of emotional well-being in protected areas, the UK's Brecon Beacons National Park offers peace, tranquility, and the quiet enjoyment of healthy exercise, inspiration, relaxation, and spiritual renewal (Brecon Breacons, 2020). The park's various amenities – health activities, fitness and leisure centers, spa hotels, and alternative therapy centers – include Peterstone Court, a country house, restaurant, and spa that takes a holistic approach to wellness by using only natural, organic products. Its other offerings include opportunities to relax in a hot tub, detox in a sauna, take a dip in the seasonal outdoor heated pool, or exercise in the gym. In the same vein, Italy's Adamello Brenta Nature Park, the largest protected area in Trentino, offers visitors different options, ranging from wellness and spa getaways in ecological hotels located in the middle of nature to excursions into the park for exclusive treatments to promote well-being (Cristina, 2020).

Meditation, yoga, and personal growth retreats are emerging in natural protected areas as well. For one, the Wellness Tree Retreat near Irton Manor, located at the southeastern end of the North York Moors National Park (NYMNP), offers yoga, meditation, swimming in the pool by candlelight, silent walks through the woods, mindfulness activities, relaxation and visualization workshops, and spa treatments with homemade organic oils. For another, the Tree Relaxation Retreat in Rosedale Abbey, ranked by *The Guardian* as one of the 25 best yoga vacations and retreats, offers Indian head massages, a bit of yoga, and spa treatments, plus walks in the moorlands and even tea and cake by the fireplace – in a short, a recipe for the perfect remedy (NYMNP, 2020).

However, if tourists seek a complete program of multi-dimensional well-being experiences, then they can consult the Renew You! program offered by Tenaya Lodge at Yosemite National Park (Tenaya Losge, 2020), which provides weekend experiences involving walks, guided mediation, talks with guest speakers, healthy cooking, relaxing visits to the spa, yoga at sunset, and, to help awaken the traveler's creative side, painting classes with wine hosted by a local artist. They also have other more specific retreat programs for women and couples, as well as stress relief getaways, each designed with the right balance of support and solitude and with the goals of re-establishing simple habits and promoting lasting change.

Last, in the spiritual dimension, volunteer work can also be a mutually rewarding experience for tourists and destinations. As stated on the official webpage of the National Parks of Finland (2020), volunteer work for the good of nature is also a great way to combine being close to nature with social relationships.

Discussion

Tourism, as a deliberate activity, is an important context for experiencing well-being (Pearce, 2009), and nature is a perfect place to get it. It is here where the three concepts interconnect, the well-being and restorative benefits of tourist experiences influence tourists' choices and subsequent attachment to destinations in nature (Voigt et al., 2011; Pyke et al., 2016). Concretely, research evidence that protected areas are crucial gateways that connect people with nature and allow them to achieve well-being through very diverse tourist experiences. Well-being experiences in nature are a growing market, a new trend in tourism, a new dynamic that responds to the current needs of consumers.

Singular and innovative experiences in protected areas have been collected to exemplify that well-being tourist experiences in nature are an emerging form of tourism in much of the world. These initiatives serve as an impetus and an example to other protected areas to continue advancing and adapting to new trends and needs of society.

As Figure 2.2 summarizes, parks and protected natural areas offer multiple benefits for society, being excellent places to develop activities for the health and well-being of tourists and residents. The landscape diversity and the ecosystem values of these areas allow, among many benefits, to develop activities that reduce stress, balance mind and body, personal growth and boost self-self-esteem, among others. In addition, protected areas allow the development of group activities in favor of inclusion and creating bonds. A total of 18 experiences, strategies and activities in natural areas have been collected, and these revolve around the multidimensional dimensions and benefits of well-being defined by the GWI (2018): environmental, physical, mental, spiritual, emotional, and, social.

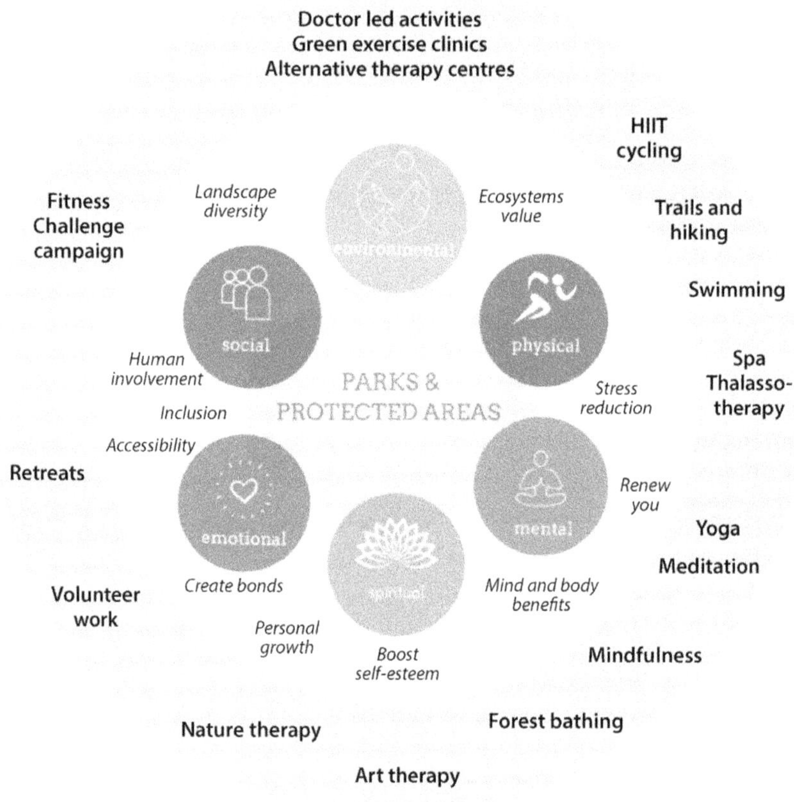

Figure 2.2: Experiences and well-being benefits of protected natural areas. Source: Authors' own elaboration

These experiences may seek a specific benefit, but by taking place in natural protected areas they end up looking, directly and indirectly, for the other dimensions. From a transversal vision, the environmental dimension of well-being is covered only by the fact of developing activities in nature, but the development of integral, holistic and well-being experiences, allows reaching for many other dimensions.

The experiences and activities described in this analysis show that countries such as the United States, Canada, Australia, and the United Kingdom have adapted and incorporated those benefits of health and both physical and mental well-being in their strategies and proposals for activities and experiences for visitors. The low presence or visibility of such experiences organized and promoted by national park systems in Europe and Central America also deserves attention, for it contrasts with the high value and potential of their many protected areas.

Managerial implications

The foregoing reflections on experiential tourism in nature as an emerging form of tourism have implications for managers, as well as indicating challenges, recommendations for improvement, and directions for future research. Above all, the careful planning and management of tourist destinations and tourism activities can allow the sector to contribute to the health and well-being of those working in tourism, locals, and travelers who seek contact with nature and memorable experiences to escape their urban realities, daily stress, and overconnectedness with technology.

Beyond that, the enjoyment of natural areas is a growing trend in tourism, and tourism management and conservation policies have to understand that protecting ecosystems is vital to sustainable tourism. Finally, the need for a new model of governance and public–private collaboration, especially in protected areas as key places for human well-being, should be noted. Managers of protected areas have an important challenge that entails identifying the potential for well-being offered by natural protected areas and co-creating appealing experiences and offerings that benefit the health of visitors and local populations while adapting to changing needs, even during global crises.

Conclusion

The search for memorable tourist experiences and connections in nature is an emerging trend that is reinforced in the current moment of change in tourist behavior. The COVID-19 pandemic has brought the relationships between environmental sustainability, economic growth, and human well-being into sharp relief, and in the search for a balance between those aspects is an opportunity to rethink the tourist model of destinations.

Protected areas offer an added value that is becoming increasingly important and with which it is very difficult to compete. Many of the activities on offer, most of which can also be performed indoors, when conducted outdoors gain the additional benefit of being in nature. Nature is a source of well-being, one that offers a wide range of opportunities and health benefits. In this way, such activities offer not only physical benefits but also mental, emotional, spiritual, and social ones.

Tourist destinations, with the collaboration of all stakeholders, might adapt their offerings to satisfy the emotional, social, and spiritual needs of today's tourists. Alliances with social, educational and health entities should be promoted to complement a program of activities of inclusive well-being in protected natural areas. Today, most protected areas are underused considering their potential, not only for the natural and ecological value of their ecosystem, but for being safe, open, well-managed and accessible for all.

The connection between well-being, experience and nature is a relationship that can yield new discoveries, and deserves to remain an object of study and analysis, especially from a multidisciplinary perspective.

Highlights

- ☐ Protected areas increasingly play a key role as a source of well-being.
- ☐ Managers of protected areas face the challenge of identifying their spaces' potential and benefits for health and well-being, and incorporating them into new tourism offerings and experiences. For that, public–private collaboration is pivotal.
- ☐ Well-being, tourism, and protected areas have a close relationship as an emergent tourist form that generates many environmental and socio-economic opportunities in the short and long term, and should be further analyzed.

Questions

1. Are the protected areas aware of the role and their contribution to the welfare of the population?
2. Why is there little presence of well-being proposals in protected areas of Europe and Latin America?

3 What planning and management instruments must be incorporated into well-being in protected areas?

Short case study: Els Ports Natural Park bet on the experiences of well-being in nature

Els Ports Natural Park (EPNP) is a mountain massif that extends beyond the south Catalonia into Aragon and Valencia. With more than 35,000 ha, the Catalan part of the Els Ports massif was declared a Natural Park in 2001, as was the Partial Nature Reserve of Les Fagedes dels Ports, which sits in the heart of these mountains. EPNP is recognized as a Special Protection Area for Birds, and as a Special Area of Conservation, included in the Natura 2000 Network. The Terres de l'Ebre region, where EPNP is located, was declared a Biosphere Reserve by UNESCO in 2013, thanks to its rich landscape, its typical Mediterranean ecosystems, and its model of biodiversity conservation (Els Ports Natural Park, 2021; Ebre Biosphere, 2021).

The villages of this protected area and its boundaries are culturally and gastronomically rich, adding to its natural features and exceptional biogeographical interest (Terres de l'Ebre Travel, 2021). Additionally, it is an unbeatable destination for activities and sports surrounded by nature, particularly hiking, cycling, and caving. Enthusiasts of botany, zoology, ethnology, geology, and hydrography can find a thousand and one reasons to make hundreds of possible itineraries (Tortosa Tourisme, 2021).

EPNP currently offers a very wide range of well-being experiences rooted in nature, an emerging form of tourism in this area. The park's natural spaces are of great environmental quality and are ideal for getting in touch with nature and for making forest baths, an experience of deep connection with nature that provides physiological, psychological, and emotional benefits (Ports Experience, 2021). EPNP's nature serves as a space to work on psychosocial well-being, and is ideal for developing emotional skills, facilitating emotional education for children and adults, and helping address problems and illnesses through introspection in nature (Nakupenda, 2021).

The nature, rurality, and diversity of this place also gives rise to authentic experiences, handmade products, and opportunities for environmental education, such as organic and nature Mediterranean cosmetics, textile craft workshops, and traditional fish farms. To date, there have already been six editions of "Els Ports, Natura & Art," a space for artistic reflection on the relationship we maintain with nature. It takes an intimate approach to the landscape with a revaluation through contemporary artistic intervention – ecological, ephemeral, and respectful – through poetry, sculpture, performance, music, etc. (Lo Pati, 2020).

As a result of all these features and experiences, the protected area is in the process of adhering to the European Charter for Sustainable Tourism. This will allow managers of protected natural areas and companies to define tourism development strategies to enhancing the park's benefits for public welfare, bringing greater visibility to them, and designing new proposals.

References

Albuquerque, H. Silva, A.M.F., Martins, F. & Costa, C (2018). Wellness tourism as a complementary activity in saltpans regeneration, in I. Azara, E. Michopoulou, F. Niccolini, F., B.D. Taff, & Clarke, A. (eds.), *Tourism, Health, Wellbeing and Protected Areas*, CABI, pp.56-68.

Ali, C. (2018) What the Heck is Forest Bathing? 5 Things You Didn't Know About Shinrin-Yoku in BC. https://www.hellobc.com/stories/what-the-heck-is-forest-bathing-5-things-you-didnt-know-about-shinrin-yoku-in-bc (Accessed: 28 October 2020).

Almeida, C., & Park, S. (2017) Exploring the tourist experience: a sequential approach, *Journal of Travel Research*, 56(1), 16–27.

Azara, I., Michopoulou, E., Niccolini, F., Taff, B.D., & Clarke, A. (eds.) (2018) *Tourism, Health, Wellbeing and Protected Areas*, CABI.

Bowler, D.E., Buyung-Ali, L.M., Knight, T.M. & Pullin, A.S. (2010) A systematic review of evidence for the added benefits to health of exposure to natural environments, *BMC Public Health*, 10(1), 456.

Brecon Breacons (2020) Brecon Breacons our National Park: Health and Well-being. https://www.breconbeacons.org/things-to-do/activities/health-and-wellbeing (Accessed: 27 December 2020).

Brent-Ritchie, J.R., Wing Sun Tung, V. & Ritchie, R. (2011) Tourism experience management research: Emergence, evolution and future directions, *International Journal of Contemporary Hospitality Management*, 23(4), 419-438.

Carballo, R., Moreno-Gil, S., León C. & Brent-Ritchie, J. R. (2015) La creación y promoción de experiencias en un destino turístico. un análisis de la investigación y necesidades de actuación, *Cuadernos de Turismo*, 35(35), 71-94.

Coelho, M.D.F., Gosling, M.D.S. & Almeida, A.S.A.D. (2018) Tourism experiences: core processes of memorable trips, *Journal of Hospitality and Tourism Management*, 37, 11–22.

Cristina, M. (2020) 10 Eco-hotels for your wellness holiday in the Adamello Brenta Natural Park, Ecobnb. https://ecobnb.com/blog/2020/11/eco-hotels-wellness-holiday-adamello-brenta-natural-park/ (Accessed 30 November 2020)

Csikszentmihalyi, M. (1975) *Beyond Boredom and Anxiety*, San Francisco, CA: Jossey-Bass.

Cvikl, D. (2016) Forest well-being tourist product: The case of shinrin-yoku, Obedska bara, Srbija. *Turisticko poslovanje*, 18, 99-107

Dutt, C. & Ninov, I. (2016) The role of mindfulness in tourism: Tourism businesses' perceptions of mindfulness in Dubai, UAE, *Tourism*, 64(1), 81-96.

Ebre Biosphere (2021). Terres de l'Ebre Biosphere Reserve. Consortium Environmental Policies. https://www.ebrebiosfera.org/ (Accessed: 23 March 2021).

Els Ports Natural Park (2021). Protection history. http://parcsnaturals.gencat.cat/ca/xarxa-de-parcs/ports/inici/ (Accessed: 23 March 2021).

Ericson, T., Kjønstad, B. G. & Barstad, A. (2014) Mindfulness and sustainability, *Ecological Economics*, 104, 73-79.

Europarc (2018). *EUROPARC Toolkit: Health & Well-being benefits from Parks & Protected Areas.* Regensburg: Europarc Federation.

Euruni, (2020). What will happen with tourism for the rest of 2020? Travel trends after the COVID-19 pandemic, 2020. https://www.euruni.edu/blog/tourism-2020-travel-tr (Accessed: 16 October 2020).

Gillezeau, N. (2020) Five easy ways to workout in nature. NSW National Parks and Wildlife Service. https://blog.nationalparks.nsw.gov.au/five-easy-ways-to-workout-in-nature/ (Accessed: 21 December 2020).

Gladwell, V.F., Brown, D.K., Wood, C., Sandercock, G.R. & Barton, J.L. (2013) The great outdoors: how a green exercise environment can benefit all, *Extreme Physiology & Medicine*, 2(3), 1-7.

GWI (2018). *Global Wellness Tourism Economy.* Miami: Global Wellness Institute.

Hernández, R. & Padrón, N. (2020) Turismo, COVID-19 y la diversificación inteligente de la economía canaria, in M R. Simancas, R. Hernández & N. Padrón (ed.), *Turismo pos-COVID-19: Reflexiones, retos y oportunidades*, La Laguna: Cátedra de Turismo CajaCanarias-Ashotel de la Universidad de La Laguna, pp. 725-735.

Highmore, B. (2002) *Everyday Life and Cultural Theory*, London: Routledge.

Japan Travel (2020) Forest bathing in Japan National Parks (2020). https://www.japan.travel/national-parks/plan-your-visit/guides-and stories/forest-bathing-in-japans-national-parks/ (Accessed: 20 November 2020)

Li, Y. (2000) Geographical consciousness and tourism experience, *Annals of Tourism Research* 27(4), 863-883.

Li, Q. (2018). *Shinrin-yoku: The art and science of forest bathing.* Penguin Life Edition.

2

Lo Pati (2020). Els Ports, Natura & Art. https://www.lopati.cat/ca/ (Accessed : 11 April 2021).

Loureiro A. & Veloso S. (2017) Green exercise, health and well-being, in G. Fleury-Bahi, E. Pol & O. Navarro (eds.), *Handbook of Environmental Psychology and Quality of Life Research*. Cham: Springer.

Maller, C, Townsend, M., Brown, P. & St. Leger, L. (2002). *Healthy Parks, Healthy People: the health benefits of contact with nature in a park context: a review of current literature*. Melbourne: Parks Victoria, Deakin University Faculty of Health & Behavioural Sciences.

Nationals Parks of Finlandia (2020) Well-being from Nature, https://www.nationalparks.fi/wellbeingfromnature (Accessed: 7 November 2020).

National Trust (2020) A beginner's guide to forest bathing. https://www.nationaltrust.org.uk/lists/a-beginners-guide-to-forest-bathing (Accessed: 23 November 2020).

Nakupenda (2021). Emotional Education. https://www.instagram.com/nakupenda.ee/ (Accessed: 10 April 2021).

Newsome, D. (2020) The collapse of tourism and its impact on wildlife tourism destinations. *Journal of Tourism Futures*, ahead-of-print.

NYMNP (2020) Health and wellbeing. North York Moors National Park Authority. https://www.northyorkmoors.org.uk/visiting/enjoy-outdoors/health-and-wellbeing (Accessed: 30 November 2020).

One Planet Sustainable Tourism Programme (2020) *One Planet Vision for a Responsible Recovery of the Tourism Sector*. UNWTO

O'Dell, T. (2007) Tourist experiences and academic junctures, *Scandinavian Journal of Hospitality and Tourism*, 7(1), 34-45.

Parks Victoria (2017) *A Guide to Healthy Parks, Healty People*. Melbourne: Parks Victoria.

Peak District (2020) Cycle Strategy. https://www.peakdistrict.gov.uk/visiting/cycle/cycle-strategy (Accessed: 7 November 2020).

Pearce, P.L. (2009) The relationship between positive psychology and tourist behavior studies, *Tourism Analysis*, 14(1), 37–48.

Pearce, P.L. & Packer, J. (2013) Minds on the move: new links from psychology to tourism, *Annals of Tourism Research*, 40, 386–411.

Pine, B.J. & Gilmore, J.H. (1999) *The Experience Economy*, Boston, MA: Harvard Business School Press.

Pope, E. (2018) Tourism and wellbeing: transforming people and places, *International Journal of Spa and Wellness*, 1(1), 69-81.

Ports Experience (2021). Forest bathing. https://www.portsexperience.com/forest-bathing/ (Accessed: 20 March 2021)

Pyke, S., Hartwell, H., Blake, A. & Hemingway, A. (2016) Exploring well-being as a tourism product resource, *Tourism Management*, 55, 94–105.

Quinlan, S. & Carmichael. B. (2010) The dimensions of the tourist experience, in M. Morgan, Lugosi, P. & Brent, R. (eds.), *The Tourism and Leisure Experience: Consumer and managerial perspectives,* Bristol: Channel View, pp 3-27.

Riley, E. (2020). What the future of adventure travel looks like, *Outside*. https://www.outsideonline.com/2413863/future-adventure-travel-after-coronavirus-pandemic (Accessed: 5 October 2020). 2

Rogan, N. (2020). 6 trends for 2021: Travel's comeback story. *Amadeus*. https://amadeus.com/en/insights/blog/six-trends-2021-travels-comeback (Accessed: 20 December 2020).

Romagosa, F., Eagles, P.J., & Lemieux, C. J. (2015) From the inside out to the outside in: Exploring the role of parks and protected areas as providers of human health and well-being, *Journal of Outdoor Recreation and Tourism*, 10, 70-77.

Russell, R., Guerry, A.D., Balvanera, P., Gould, R.K., Basurto, X., Chan, K.M.A., Klain, S., Levine, J. and Tam, J. (2013) 'Humans and nature: How knowing and experiencing nature affect well-being', *Annual Review of Environment and Resources*, 38, 473-502.

Ryan, R. M., Weinstein, N., Bernstein, J., Warren Brown, K. Mistretta, L., & Gagné, M. (2010). Vitalizing effects of being outdoors and in nature, *Journal of Environmental Psychology*, 30, 159-168.

Ryse, R.F. & Meers, J.E. (2012) EcoWellness: The missing factor in holistic wellness models, *Journal of Counseling & Development*, 90, 400-406.

Same S. & Larimo J. (2012) *Marketing theory: Experience marketing and experiential marketing*, paper presented at the 7th International Scientific Conference Business and Management, 10-11 May, Vilnius.

Selstad, L. (2007) The social anthropology of the tourist experience. Exploring the 'Middle Role', *Scandinavian Journal of Hospitality and Tourism*, 7(1), 1933.

Selvans (2020) Banys de bosc. https://selvans.coop/categoria/projectes/banys-de-bosc/ (Accessed: 30 November 2020).

Smith, M. K. & Puckó, L. (eds.) (2016). *The Routledge Handbook of Health Tourism*. Abingdon: Routledge.

Sugathan, P. & Ranjan, K. R. (2019) Co-creating the tourism experience, *Journal of Business Research*, 100, 207-217.

Tenaya Lodge (2020) Yosemite Wellness Retreat Weekends, https://www.tenayalodge.com/spa/spa-retreats (Accessed: 28 October 2020).

Terres de l'Ebre Travel (2020). Terres de l'Ebre Travel. https://terresdelebre.travel/en (Accessed: 3 April 2021).

Thellier, I. (2011) Esprit de Picardie: le storytelling au service d'une marque expérientielle et relationnelle, *Revue Espaces Tourisme et Loisirs*, Décembre, 32-37.

Tortosa Toursime (2021). Between Two Natural Parks. https://www.tortosaturisme.cat/en/ (Accessed: 29 March 2021).

Tripadvisor (2020) Beyond COVID-19: The road to recovery for the travel industry. https://www.tripadvisor.com/Covid19WhitepaperMay2020 (Accessed on 10 September 2020)

UNWTO (2020) *Global Guidelines to Restart Tourism*, Madrid: UNWTO.

UNWTO (2021) *COVID-19 and Tourism | 2020: A year in review*, eLibrary: UNWTO.

USNPS (2020) Benefits of Hiking. National Park Service – Department of Interior United States https://www.nps.gov/subjects/trails/benefits-of-hiking.htm (Accessed: 3 November 2020).

Vada, S., Prentice, C., & Hsiao, A. (2019). The influence of tourism experience and well-being on place attachment, *Journal of Retailing and Consumer Services*, 47, 322–330.

Visit Victoria (2020) Forest Bathing. https://www.visitvictoria.com/Features/Forest-bathing (Accessed: 30 November 2020).

Voigt, C., Brown, G. & Howat, G. (2011) Wellness tourists: in search of transformation, *Tourism Review*, 66(1/2), 16–30.

WWW-UK (2020) 5 ways to connect to nature to help our wellbeing. https://www.wwf.org.uk/5-ways-connect-nature-help-our-wellbeing (Accessed: 22 October 2020).

3 New clues for tourism planning and management from consumer neuroscience

Carla Fraga and Airton Rodrigues

Abstract

The chapter's primary goal is to look at the future of tourism and discuss consumer neuroscience in the context of tourism planning and management. We first provide theoretical and concept views about the theme of neuroscience in tourism. Then, we describe the different sensors and devices that make it possible to measure and understand consumers' emotional responses. Following, we show the importance of consumer neuroscience to tourism planning and management while facing the 21st century's challenges. Through neuroscience, it is possible to understand cognitive and emotional processes inaccessible to traditional research. This chapter contributes from a bibliographic approach with a context of emerging dynamics in tourism and hospitality.

Keywords

Tourism; neuroscience; consumer; planning; management.

Introduction

Relationships between neurosciences and other disciplines give rise to new scientific knowledge: neuromarketing, neuroeconomics, and neuropsychology. Karmarkar and Plassmann (2019) delivered an overview of the growing field of consumer neuroscience. Tourism is a complex phenomenon that involves consumption and the paradigmatic view for its treatment is the General System Theory (Lohmann & Panosso Netto, 2008). Nevertheless, according to Tribe (1997) tourism has two fields, a business and a non-business field, while tourism studies draw four approaches from the world of thought (multidisciplinary, general interdisciplinarity, business interdisciplinary) and the world of practice (extra disciplinarity). It is worth noting that the meanings and senses of tourism experience as an expression of consumption have undergone transformations during the 21st century, be it in the context of combating overtourism as witnessed before the COVID-19 pandemic, or in creating development strategies for tourism in a post-pandemic era, since confidence for travel has been shaken by the fear experienced through the pandemic (see Mayer & Coelho, 2021).

It is necessary to recuperate from social distancing impositions brought on by the pandemic. Access to destinations through virtuality in detriment of physical dislocation seems to have enhanced causing tourism consumption to transform, employing the intermediation of disruptive technologies, notedly those that use the global computer network through the internet, for example, virtual tours with the aid of augmented reality goggles. This has been made possible since the advancements in Web 1.0, 2.0, 3.0, 4.0 (Latorre, 2018) and the internet of things (Ashton, 2009) which has become more and more useful for those with access to the internet. Therefore, tourism consumption and tourists' and stakeholders' behaviours are being transformed, making emotions the new frontier for tourism studies, as posited by Moyle et al. (2019).

This troubled and uncertain environment accentuates the difficulty in business planning and management and brings more uncertainties to the different agents present in the tourism value chain. As observed, changes were not restricted to tourist behaviour; on the contrary, they have also impacted the business environment and the new landscape that appears along the value chain. One of the most impactful changes is the one resulting from mobile technologies, which have established a new enterprise-consumer relationship (Guerrero-Rodríguez et al., 2020). For example, a report of Euromonitor International (Bremner, 2020) showed the knockout suffered by tourism as a result of the COVID-19 pandemic, suggesting that the industry is expected to better recover and advance through digital media, compared to other segments in the value chain. It is noteworthy that eWOM (electronic word of mouth) through social media exerts a strong influence on tourist behaviour and travel intention. eWOM affects cognitive evaluations by consumers about available brands and their characteristics and helps to form beliefs that result in the choosing of a desirable option (Pourfakhimi et al., 2020). 3

Tourism dynamics demand the discovery of new theoretical and methodological perspectives for planning and management, and it is undeniable that the advances in the comprehension of consumers' emotional responses appear as an epistemological frontier for the understanding of the tourism phenomenon itself, in the present or in the future. As an example, Al-Kwifi (2015) studied brain activation from attractive images for tourists by using functional magnetic resonance (FmRI), revealing the technological impact that can be made by neuroscience's tools in planning and managing tourism destinations. However, it is important to stay vigilant, as on the other side the "*... the field of consumer neuroscience has been plagued with pseudoscience and 'neurohype' and researchers have experienced some disappointments when trying to incorporate these measures into their research*" (Niedziela & Amborze, in press).

From this perspective, theories, methods, and tools from neuroscience can contribute to neuro-tourism as an emergent topic in tourism research, with neuro-tourism becoming a discipline itself. This chapter's primary goal is to look at the future of tourism and discuss consumer neuroscience in tourism planning and management. The first section provides theoretical and conceptual views about the thematic. Then, we describe the different sensors and devices that measure and understand consumers' emotional responses. Following, the third section of this chapter shows the importance of consumer neuroscience to tourism planning and management, in the face of the challenges of the 21st century, before conclusions are drawn and implications are discussed.

Consumer neuroscience and tourism: theoretical and concept approaches

It is not only recent history that has inflicted significant changes to the tourism phenomenon. Since after World War II, a shift in focus has been observed from the delivery of tourism products to tourist experiences, a new management paradigm that emphasizes mass customization and the experience's design as a frontier to be reached by organisations (Brent Ritchie et al., 2011). Passive consumption gave place to a context of active search for novelty and pleasant, immersive and transformative experiences (Taheri et al., 2019). The processes of choosing and deciding between different offers are complex as they require to build a strong relationship with a 'future me' – a projective image of the experience to be lived with a transformational perspective leading to a process of high involvement, both psychological and financial (Ramsøy et al., 2019). Experiences, whether ordinary or extraordinary, transform lives and reconstitute reality through fantasy, feelings and enjoyment. In this way, tourism as a business is ultimately responsible for this mental construction (Agapito et al., 2013). The COVID-19 pandemic caused even more profound impacts, presenting a necessity for developing new tourism products and adjusting existing ones to allow companies to adapt to a new competitive scenery (Smykova et al., 2020).

The design of tourist experiences is related to a quest for intrinsic rewards, which may translate into positive states of mind and gratification (Mannell & Iso-Ahola, 1987). This effect is understood as a pursuit of a psychological process of well-being. Comprehension of this process is intimately related to research methods that make it possible to understand the experience's anatomy, how it can be organized, its duration, its intensity and above all its memorability (Mannell & Iso-Ahola, 1987).

Therefore, a critical step for developing this planning and management paradigm is the profound realization of tourist experiences through research processes that allow one to explore tourists' minds more deeply (Brent Ritchie et al., 2011). On the one hand, any effort of delivering an experience is conditioned to the adequate shape of the offers of tourism products available. On the other hand, there is the need to assess the effectiveness (and efficiency) of marketing efforts focused on promoting tourist destinations (Bastiaansen et al., 2018).

Neuroscience, particularly consumer neuroscience, may prove to be a powerful resource to comprehend how the mind processes certain stimuli related to tourism and to, consequently, quantify identified emotional processes which arise from different tourist experiences. Emotions and all that is related to them, such as feelings, memory, attention, cognition and behaviour, have always incited great curiosity, and the study of the nervous system is just a part of a series of disciplines prior to what is known nowadays as neurosciences, where discoveries have been made from philosophy, medicine and biology among other disciplines, which have also shaped the beginnings of neuroscience (Rooney, 2018).

However, from neuroscience's modern history it is noticeable that there are several levels of analyses, namely molecular, cellular, structural, behavioural and cognitive (see Bear et al., 2017); these levels ought to be considered in tourism studies, since the tourism phenomenon is about experiences that are processed from stimuli, senses and emotions, embracing, for example, attentional availability, feelings and memorabilia.

Tourism, as one of the applied social sciences, has in almost no way aligned with neuroscience's advances for knowledge generation, albeit Scott (2020) has already affirmed that in the next 75 years, neurosciences and cognitive psychology will play a fundamental role for a new paradigm in the treatment of tourism. This implies that in theoretical and methodological terms, there will be progress in the direction of experiment studies. Viglia and Dolnicar (2020:2) highlight that *"of 88 articles published in* Annals of Tourism Research *in 2018, for example, only seven (less than 8%) were based on experiments"*. These authors give a fundamental contribution to the development of experimental research in tourism and hospitality, demonstrating types of experiments and presenting an 'experimental design checklist' (Viglia & Dolnicar, 2002) in social sciences.

Although not every experiment made in tourism and hospitality is based in neuroscientific knowledge, this knowledge relied on the scientific process (observation, replication, interpretation and validation) and is anchored in scientific method. It thereby unites scientists from various areas in search of new findings about the nervous system, as suggested by Bear et al. (2017). Therefore, from Viglia and Dolnicar's (2020) observations it is possible to frame tasks, examples and aspects to consider based on neuroscientific knowledge in general.

Specifically, as tourism is a phenomenon which involves the consumption of products, services, experiences, images etc. some developments of consumer neuroscience, as a part of the neurosciences, can very well contribute with new clues on a theoretical and methodological basis. It is important to differentiate consumer neuroscience from neuromarketing. Ramsøy (2019) defines neuromarketing as the commercial use of neuroscience's tools and insights, whilst consumer neuroscience is responsible for basic research regarding how consumption decisions are made. The following section addresses the sensors and devices used to measure consumer's emotional responses through a practical approach.

The role of sensors and devices

Before presenting and discussing sensors and devices, it should be noted that a successful marketing strategy in tourism must be based on the ability to activate emotional elements in consumers and to operationalize, through different communication channels, the expected projection of the 'future me'. However, this is a challenge in terms of measurement (Bastiaansen et al., 2018). For example, marketing researchers have been mostly skeptical about the usage of behavioural measurements obtained by self-reports due to the limitations in providing an efficient dimension of the internal reaction to external stimuli (Wang & Minor, 2008).

Habitually people are requested to reconstruct their experiences by memory and to narrate their emotional experiences in questionnaires and interviews (Guerrero-Rodríguez et al., 2020). However, this form of comprehension of consumer's experience may have serious limitations (Ma et al., 2014). First, given the nature of memory which holds selective processes, making it very likely that the information given is filtered in an unconscious and involuntary manner. As well as the possibility that the interviewed subjects make deliberate reconstitutions of their experiences, not narrating facts in a faithful way (Ma et al., 2014).

This aspect brings uncertainties to information construction regarding consumers and may bring risks to decision-making processes in management. Thus, it is necessary that tourism's planning and management consider more accurate, wide-ranging and impartial measurements of psychological processes in order to reflect a wider and deeper understanding of the mental processes which influence consumers' decision-making processes (Wang & Minor, 2008). Therefore, psychophysiological type measurements can aid to more profoundly understand aspects that consumers may try to hide, may not know or can't verbalize in traditional interviews. This, in itself, incites a series of ethical questions in regards to consumption that should be clarified when these methods are employed.

It is relevant to mention that a variety of psychophysiological techniques have been used for measuring consumer reactions to marketing stimuli since the 1960s, and that they allow for the identification of various cognitive and emotional processes which take place in the brain (Wang & Minor, 2008). The most common measures come from the central nervous system (CNS) that includes an analysis of non-hemispheric brain waves, hemispheric lateralization and analysis of brain images; from the autonomous nervous system (ANS) referring to pupil response, electrodermal analysis, voice tone analysis, cardiac frequency response and vascular activity; and from the somatic nervous system (SNS) which involves facial muscle activity and analysis of eye movement (Wang & Minor, 2008).

Psychophysiology tries to establish relations between psychological, social and behavioural phenomena and physiological reactions that are not subjected to a person's voluntary control (Cacioppo et al., 2007). One of these measurements is Galvanic Skin Response (GSR), facial and abdominal muscle activity, heart rate variability and eye-tracking. These can be more efficient than self-report based measurements since they capture reactions while they happen in a discreet way and with no associated cognitive bias regarding the memory effect and inconsistencies in the cognitive process (Guerrero-Rodríguez et al., 2020).

It's vital to consider the limitations of traditional researches, whilst also recognizing their relevance. Kahneman (2011) shows that decision-making happens fundamentally on two bases: System 1 and System 2. The first, sustained in automatic and emotional processes, guides the consumer's impulsive behaviour and rules sentimental attitudes. The second, complex and deliberative processes, guides one's rational behaviour. Choices based on memory tend to reflect more the processing of System 1 (due to the dominance of emotional aspects) while choices based on stimuli (e.g. promotions) reflect relatively more the processing of System 2, anchored in rationalization (Rottenstreich et al., 2007).

This duality calls for market studies that observe both of these systems, and for a holistic perspective of how decision-making occurs in face of a wide range of possibilities of places and experiences. While traditional market research studies System 2, consumption neuroscience, through psychophysiology allows for the comprehension of System 1's functioning. Table 3.1 presents descriptions and observations about sensors and devices common and useful in the study of System 1. In this table, PET and fMRI weren't included due to the difficulty in access when it comes to traditional market research (because of cost and presentation methods).

Table 3.1: Sensors and Devices

Facial Coding

Facial Coding is an important resource to evaluate an individual's emotional responses to specific environmental stimuli. The Facial Action Coding System (Cohn & Ekman, 2005) uses computer image analysis to process measures of facial symmetry and through these results produce inferences regarding emotional states of individuals. This is a widely used resource because of the intuitivity of its results and because it is not necessary to connect devices on participants.

EMG

Another face monitoring resource is Facial Electromyography (EMG). The EMG utilizes sensors positioned in specific points in the subject's face to measure activity on key muscular groups associated with emotional reactions (Spinelli & Niedziela, 2016). This measurement is used to identify the value of responses of affection and external stimuli (Wang & Minor, 2008). It allows for a highly precise identification of emotional states, not possible in a traditional interview and, through computational resources, enables the synchronization between stimuli and landscapes, itineraries or product proposals (experiences to be offered) and to quantify the favourable or unfavourable responses in face of these stimuli.

GSR

A widely used resource to assess levels of stimuli impact related to tourism is called Galvanic Skin Response (GSR). GSR is used to measure electrodermal activity, a property of the human body that causes continuous variation of the skin's electrical properties (Spinelli & Niedziela, 2016). The range of conductance is measured in microSiemens (μS) and the response peak is associated to bigger emotional arousal – this resource may be utilized, for example, when a movie campaign is launched for a specific public and, through synchronicity between the emotional peak caught in scanning and the film's part which caused it, it is possible to identify the biggest emotional responses without identifying the value of those answers (Guerrero-Rodríguez et al., 2020). Psychophysiology shows that sweat glands can reflect psychological activity and, therefore, that changes in electrodermal activity within the sympathetic nervous system may be a result of interest, excitement or pleasure (Wang & Minor, 2008). However, it is always important that the nature of this value is obtained through interviews since the GSR is capable of measuring the intensity but not the emotional sense of answers.

EEG

Electroencephalogram (EEG) is used more frequently for brain waves mapping in marketing research (Wang & Minor, 2008). It evaluates variations in the brain's electrical activity frequency (Young, 2002). The EEG allows participants to see stimuli in a relatively natural way, while they register the brain's reactions at a highly elevated speed (milliseconds) with great temporal resolution, registering mental actions at the exact moment they occur (Bastiaansen et al., 2018). Through this device, it is possible to put tourists before landscapes and destinations and assess cognitive responses, optimizing marketing efforts. One possibility for analysis is the examination of the difference in cerebral activity in each hemisphere, left and right, which allows for the identification of emotional responses when individuals are exposed to stimuli such as communication proposals of tourist packages or scenes from itineraries (Bastiaansen et al., 2018).

Eye-tracking

The eye's movements are continuous and involuntary and the processes of fixation (brief stops, in a rate of milliseconds, during visual scanning to absorb a greater quantity of information from a visual scene) show cognitive interest areas (Martinez-Conde et al., 2004). Eye-tracking equipment enables one to monitor and trace visual scanning and is one of the primary methods used to comprehend System 1 responses. This is due to the fact that it allows for the quantification of visual attention, as it monitors where people see visual messages (e.g. advertisements), verbal messages (e.g. price discounts) or physical messages (e.g. products on shelves) (Boz et al., 2017). Eye-tracking is able to monitor the eye's direction, fixation points (momentary stops in the process of visual scanning) and pupil dilation usually associated with emotional engagement (Spinelli & Niedziela, 2016).

Although Table 3.1 doesn't exhaust sensors and devices possibilities, it offers an important overview to developments in tourism studies. Generally, it is observed that each monitored element is a gateway to a specific sensory information (face, skin, eyes etc). Thereby, paying attention to the sensory dimension in tourism experiences is essential since: (a) human senses are crucial for one's perception of the world; (b) sensory stimuli influence consumer behaviour; and (c) places and environments such as destinations are multi-sensory, providing multi-sensory encounters (Agapito et al., 2013). Beyond the five human senses (sight, hearing, smell, taste and touch), which give information about external environments (exteroceptive senses), neuroscience proposes that sensory signals received by the brain can also come from inside the body, those belonging to body consciousness (interoceptive senses) (Agapito et al., 2013).

The use of psychophysiological methods to study people's emotional responses in propaganda, marketing and media is recent but its use in tourism research is relatively new and has been raising more interest and getting more acceptance among scholars and researchers (Guerrero-Rodríguez et al., 2020). Thus, the next section discusses tourism planning and management from consumer neuroscience aspects.

New clues to tourism planning and management

As mentioned above, over the course of the last decades, tourism underwent profound transformations as a result of the birth of new travelling practices, technologies and digitalization processes (Guerrero-Rodríguez et al., 2020). Consumers' projection of self-image in the destination plays now a central role in a tourism decision-making process and is the fundamental building block in marketing research (Bastiaansen et al., 2018), which must be developed in a multidisciplinary effort, along with areas such as sociology, psychology and organizational behaviour among others (Taheri et al., 2019). However, instruments such as market research, essential to tourism planning and management begin to find limitations in studies when constrained to traditional methodologies.

Neuroscience's tools can obtain the human body's direct neurophysiological data as observed in Table 3.1, serving as potential measures of some constructs of happiness or satisfaction regarding tourism and regarding the intention of revisiting (Ma et al., 2014). Thus, it is vital to remember that neuroscience analyses the nervous system to comprehend the biological bases of behaviour (Squire, 2008). The idea is that the central value of tourism, as consumption, is in offering tourist experiences and the construction of memory from them; and neurophysiological measurements can help quantify and characterize the quality of those experiences and memories in order to better utilize marketing efforts. The study of consumers in the light of neuroscience's tools and methods is enhancing understanding on the formation and choosing of preferences (Ramsøy et al., 2019). Therefore, anchored in what has been demonstrated in the previous sections, the way this knowledge born in the neurosciences can aid in tourism planning and management becomes a focus of discussion.

On the side of tourism planning it is known that it involves different steps (analysis of the internal and external environment, diagnosis, prognosis, strategy, implementation and evaluation), as was summarized by Lohmann and Panosso Netto (2008). Thus, it is important to notice that the alignment between neuroscience and studies regarding the neural mechanism subjacent to tourist behaviour, gave birth to an area of studies called neuro-tourism (Ma et al., 2014). For example, regarding the complexities of tourism planning and management, a main one is the presence of expectations before travel, which need to be fulfilled so that the tourist feels satisfied and becomes loyal to the destination whilst advertising it positively. Therefore, knowing tourists' emotional responses can enhance tourism comprehension to a new level. In practical terms, destinations can better prepare their positioning. Al-Kwifi's (2015, p.188) study, which considers consumer neuroscience, highlights:

> *"If the real experience matches the expectation, tourist satisfaction is achieved, which in turn influences the subsequent evaluation of the vacation and future intentions to revisit it. The positive experience deriving from positive images about the destination would lead to a pleasant evaluation of a destination, causing destination loyalty".*

The tourist experience can be seen as a continuous process of perception of different elements which tourists live through (pre-trip, during the trip, post-trip), involving cognitive processing of the complex external environment, as much as the emotional experience (Ma et al., 2014). Some examples of products (trips) and the kind of research that can be carried out to better understand how System 1 responds to these stimuli are evidenced in the literature by Smykova et al. (2020) as following:

- In a proposition of a tour's itineraries, utilize resources from neuroscience to evaluate the perception of natural resources, landscapes and sceneries and the perception of activities;
- In visits to beaches, evaluate subjacent reactions to aquatic resources, beach sceneries and architecture/design of hotels;
- In the case of spas and itinerary offers related to wellness, evaluate and measure the effectiveness of these experiences in provoking relaxation and reducing stress;
- Neuroscience can help assess the brain's reaction and the quality of the experience in radical situations such as skydiving, paragliding and other activities of this genre.

The complexity of tourism planning and management involves perceiving that it is composed of internal and external factors which ought to be considered. For example, Agapito et al. (2013) explain that one side is composed by external facts, some of which are under the control of companies that comprehend the environment where tourists are placed along with numerous environmental variants with which it interacts. On the other side there are internal factors, where sentimental and cognitive processes work in a complex and instantaneous way, guiding behavioural responses. Thereby, understanding the mechanisms which involve internal factors is imperative (Agapito et al., 2013). Consequently, from an epistemological perspective, the role that other areas of expertise (each with its own theories and methods) may play in order to compose the repertoire of tourism planning and management, becomes a constant requirement for the development of new knowledge about tourism and the resolution of problems, both in the field of commercial and non-commercial tourism (see the model by Tribe, 1997).

Considering the fact that consumer neuroscience relates to market research, across their experimental methodologies, it is possible to identify and characterize perceptions of value and differential from diagnosis (notably with real and potential demand) that are useful for the design of prognosis, and in turn, are key for the implementation of tourist destinations' positioning strategies. Moreover, monitoring strategies aimed at checking the course of development in tourism against expected results can also benefit from consumer neuroscience's studies.

Furthermore, neuromanagement is already a reality, as it is understood as a relatively mature discipline, coined by Ma in 2006, which intends to explore human cerebral activities in face of various management and financial problems, by using neuroscience's cognitive tools such as electroencephalogram, facial electromyography, positron emission tomography, and functional magnetic resonance imaging (Ma et al., 2014). To this end, to look at epistemological and marketing developments in tourism seems to be essentially looking at neuroscience's advances, particularly in consumer neuroscience.

Conclusions

This chapter's aim was to identify new clues for the planning and management of tourism from neuroscience, notably consumer neuroscience. In this sense, an epistemological progress in course has been identified which designates the construction of a new paradigm for dealing with tourism, one based in neuro-tourism as a discipline in itself. Some limitations belonging to the interpretation of consumer neurosciences applied to planning and management of tourism are: (1) utilizing these types of sensors and devices is still expensive; and (2) personnel with multidisciplinary and interdisciplinary knowledge are required to put in practice theories and methodologies that really does not form part of a pseudoscience or a 'neurohype'.

This argument must be put under question, based on the models by Tribe (1997) and others regarding the construction of knowledge in tourism. As an example, in the Jafari Model (Lohmann & Panosso Netto, 2008, p.43) the Centre for Tourism Studies considers departments (or disciplines) that compose the course such as administration (which involves management of tourism organisations) and therefore neuromanagement – making use of neuroscientific resources to comprehend subjacent processes that happen in the mind and that can be applied to the planning and management of tourism, opening up the possibility to revisit the very composition of the course departing from the Centre of Tourism Studies. In contrast, in terms of temporal space, it is possible to comprehend that challenges are numerous since the beginning of the 21st century. As a result of the COVID-19 pandemic, social distancing was introduced as a safety measure, hence inhibiting travel of all natures including for tourism. This has become one of the biggest challenges for the planning and management of tourism.

Finally, the importance of research ethics is reaffirmed, regarding the submission of experiments to ethics committees and the rigorous following of current legislations. The importance of transparency in research, the reliability of data, scientific reproducibility and a responsible conduction of data analysis and communication, are highlighted. These are essential elements for the development of a relationship between neurosciences and consumption in the context of the planning and management of tourism.

Highlights

- ☐ Neuroscience, particularly consumer neuroscience, may prove to be a powerful resource to comprehend how the mind processes certain stimuli related to tourism consumption.
- ☐ The importance of consumer neuroscience to tourism planning and management increases, facing the challenges of the 21st century.

Questions

1. Why is it important to comprehend tourists' emotional responses? How can consumer neurosciences be of help in achieving that?
2. Choose a sensor/device described in this chapter and explain how it could help in tourism planning and management.
3. Identify one limitation found when using consumer neurosciences principles in planning and managingof tourism. Explain how this limitation can be mitigated.

References

Agapito, D., Mendes, J. & Valle, P. (2013) Exploring the conceptualization of the sensory dimension of tourist experiences. *Journal of Destination Marketing & Management* 2(2), 62–73.

Al-Kwifi, O.S. (2015) The impact of destination images on tourists' decision making: A technological exploratory study using fMRI. *Journal of Hospitality and Tourism Technology*, 6(2), 174-194.

Ashton, K. (2009) That 'Internet of Things' Thing. June 22, 2009. *RFID Journal*. http://www.itrco.jp/libraries/RFIDjournal-That%20Internet%20of%20 Things%20Thing.pdf

Bastiaansen, M., Straatman, S., Driessen, E., Mitas, O., Stekelenburg, J. & Wang, L. (2018) My destination in your brain: A novel neuromarketing approach for evaluating the effectiveness of destination marketing. *Journal of Destination Marketing & Management* 7, 76–88.

Bear, M.F., Connors, B.W. & Paradiso, M.A. (2017). *Neurociências: desenvendando o sistema nervoso*, 4nd. Porto Alegre: Artmed.

Boz, H., Arslan, A. & Koc, E. (2017) Neuromarketing aspect of tourısm pricing psychology, *Tourism Management Perspectives*, 23, 119–128.

Bremmer, C. (2020). Accelerating Travel Innovation after Coronavirus. Euromonitor International.

Brent Ritchie, J. R., Wing Sun Tung, V. & J.B. Ritchie, R. (2011) Tourism experience management research, I*nternational Journal Contemporary Hospitality Managment*, 23(4), 419–438.

Cacioppo, J. T., Tassinary, L. G. & Berntson, G. G. (Eds.) (2007). *Handbook of Psychophysiology* (3rd ed.). Cambridge University Press.

Cohn, J. F. & Ekman, P. (2005) *Measuring facial action* in J. A. Harrigan, R. Rosenthal, & K. R. Scherer (Eds.) *Series in Affective Science. The new handbook of methods in nonverbal behavior research*. Oxford University Press, 9–64.

Guerrero-Rodríguez, R., Stepchenkova, S. & Kirilenko, A. (2020) Experimental investigation of the impact of a destination promotional video with physiological and self-reported measures. *Tourism Management Perspectives* 33.

Kahneman, D. (2011). *Thinking, Fast and Slow*. New York: Farrar, Straus and Giroux.

Karmarkar, U. R., & Plassmann, H. (2019). Consumer Neuroscience: Past, Present, and Future. *Organizational Research Methods*, 22(1), 174–195.

Latorre, M. (2018). *História de las Web 1.0,2.0,3.0 y 4.0*. Universidad Marcelino Champagnat. https://bit.ly/3bVydO3

Lohmann, G. & Panosso Netto, A. (2008) *Teoria do turismo: conceitos, modelos e sistemas*. São Paulo: Aleph.

Ma, Q. G., Hu, L. F., Pei, G. X., Ren, P.Y. & Ge, P. (2014) Applying neuroscience to tourism management: a primary exploration of neurotourism. *Applied Mechanics and Materials* 670-671, 1637–1640.

Mannell, R. C. & Iso-Ahola, S. E. (1987) Psychological nature of leisure and tourism experience. *Annals of Tourism Research* 14(3), 314–331.

Martinez-Conde, S., Macknik, S. L. & Hubel, D. H. (2004) The role of fixational eye movements in visual perception, *Nature reviews. Neuroscience* 5(3), 229–240.

Mayer, V.F. & Coelho, M. de F. (2021) Sonhos interrompidos: memórias e emoções de experiências de viagem durante a propagação da Covid-19. *Revista Brasileira de Pesquisa em Turismo*, 15(1), 1-20.

Moyle, B.D., Moyle, C., Bec, A. & Scott, N. (2019) The next frontier in tourism emotion research. *Current Issues in Tourism*, 22(12), 1393-1399.

Niedziela, M.M. & Ambroze, K. (in press) *Food Quality and Preference*, https://doi.org/10.1016/j.foodqual.2020.104124

Pourfakhimi, S., Duncan, T. & Coetzee, W. J.L. (2020) Electronic word of mouth in tourism and hospitality consumer behaviour: state of the art. *Tourism Review* 75(4), 637–661.

Ramsøy, T. Z. (2019) Building a Foundation for neuromarketing and consumer neuroscience research, *Journal of Advertising Research*, 59(3), 281–294.

Ramsøy, T.Z., Michael, N. & Michael, I. (2019) A consumer neuroscience study of conscious and subconscious destination preference, *Scientific Reports* 9(1).

Rooney, A. (2018) *A história da neurociência.* São Paulo: M. Books do Brasil Editora Ltda.

Rottenstreich, Y., Sood, S. and Brenner, L. (2007) Feeling and thinking in memory-based versus stimulus-based choices. *Journal of Consumer Research,* 33(4), 461-469.

Scott, N. (2020) Cognitive psychology and tourism – surfing the 'cognitive wave': a perspective article, *Tourism Review*, 75(1), 49-51.

Smykova, M., Kazybaeva, A. & Tkhorikov, B. (Ed.) (2020) Neuromarketing Approach to Assessing Tourism Products. *Contemporary Problems in the Development of Economic, Financial and Credit Systems,* 9th-11th September. Federal State Autonomous Educational Institution of Higher Education. Belgorod, Russia.

Spinelli, S. & Niedziela, M. (2016) Emotion measurements and application to product and packaging development, in P. Burgess, (Ed.) *Integrating the packaging and product experience in food and beverages. A road-map to consumer satisfaction.* Duxford UK: Woodhead Publishing, 77–119.

Taheri, B., Hosany, S. & Altinay, L. (2019) Consumer engagement in the tourism industry: new trends and implications for research, *Service Industries Journal* 39(7-8), 463–468.

Tribe, J. (1997) The indiscipline of tourism. *Annals of Tourism Research,* 24(3), 638-657.

Viglia, G. & Dolnicar, S. (2020) A review of experiments in tourism and hospitality. *Annals of Tourism Research,* 80. 1-15.

Wang, Y. J. & Minor, M. S. (2008) Validity, reliability, and applicability of psychophysiological techniques in marketing research. *Psychology & Marketing,* 25(2), 197–232.

4 Air connectivity for leisure tourism: the way forward

Pavlos Arvanitis

Abstract

International tourism arrivals by air account for almost 60% of all tourism arrivals. Despite the spectacular increase in air connectivity there are distinct differences between connectivity types, with direct and indirect connectivity, airport connectivity and hub connectivity. Tourism destinations are impacted by the air connectivity type that is available in their nearby or serving airport. The aim of this chapter is to discuss the tourism dynamics resulting from the emerging transformations in air connectivity, and the implications on international air travel. Increased air connectivity is linked to economic growth and development, however direct air connectivity has been driving both tourism and air transport industries. The introduction of new aircraft which can fly longer and more economically is likely to transform direct connectivity and hub connectivity at the same time. Airline business models have evolved over the last 20 years and it is highly likely that this transformation will continue to unfold since the market and the passengers' needs are constantly evolving. Implications for airlines and destinations will be discussed, outlining the trends which are dominating the industry in terms of connectivity and its relation to tourism destinations.

Keywords:

Air connectivity, tourism destinations, air transport, tourism

Introduction

The aim of this chapter is to discuss the connectivity issues that have developed over the last decades following the boost of air traffic movements and the relationship with tourism. There is reference to airline business models and connectivity models, followed by proposals on the future development of connectivity patterns for air traffic passengers. The latest developments, related to the global pandemic make predictions more difficult, however, technological improvements and changes in the passengers' behaviour might affect the future decision making of airlines as well as the shape of the connectivity networks.

Connectivity networks have existed for many centuries; originally known as trade routes both by sea and land, then they expanded to rail networks followed by road networks and finally by air connectivity networks (Morrison & Winston, 1995; Guimera et al., 2005; Cheung et al., 2020). The common characteristic of these networks is that they all need a significant and substantial degree of infrastructure to make the network feasible. In the beginning it had to do with the size of the ships and the ports, followed by the size and carrying capacity of the rail engine and the carriages and connectivity points of the railway tracks and railway stations. When automobiles were invented, the width of the roads and the load capacity of the vehicles determined the volume of people and goods that could be carried. Over the last hundred years, with the evolution of commercial air transport, connectivity networks are determined by the airport size, the airways' capacity and aircraft size.

The common denominator in the development of all the above networks has been trade, which has been the driving force. Since the introduction of the railway, people started travelling more for leisure purposes, which led to the development of tourism as it became further affordable. The massive expansion of tourism is linked with the introduction of air services where high volumes of leisure passengers travelled domestically or internationally. In 2018, international tourist arrivals exceeded 1.4 billion (UNWTO, 2019a) out of which 58% travelled by air, 37% by road and the remaining 2% by train and 4% by water. In 2000, the share of air travel was 46% and road 58% which reflects the increasing trend for air transport demand and connectivity. In 2019, UNWTO published a press release stating that:

> *"the stability of fuel prices tends to translate into affordable air travel while air connectivity continues to improve in many destinations, facilitating the diversification of source markets. Trends also show strong outbound travel from emerging markets, especially India and Russia but also from smaller Asian and Arab source markets"* (UNWTO, 2019b).

Despite extensive research in the area of air connectivity, there is limited research in a combined approach taking into consideration both air transport and tourism related (leisure) connectivity. It is widely accepted that air connectivity results in the financial prosperity of regions and areas that are served by direct international flights. This chapter aims to address the new challenges and applications in air connectivity between tourism generating and tourism receiving destinations around the world. Given that air connectivity is crucial for leisure tourism, the contribution of this chapter mainly covers the area of mobilities in tourism, and partially elements of destinations' competitive advantage and global travel patterns. Mobility has been associated with components of geography, place and space (Coles & Hall, 2006) although emphasis has been mostly on branding and marketing at destination level. This chapter highlights that despite socio-political boundaries that dictate human mobility across regions, territories and countries, technological developments and enhancements, facilitate human mobility beyond the aforementioned boundaries (Coles, et al., 2004; Hall et al., 2019). There are approaches, including this chapter, which cross disciplinary boundaries, as well as more traditional geographical perspectives on tourism and mobility (Coles & Hall, 2006). New aircraft technology in line with developments in marketing, distribution, accessibility, visa restrictions and unexpected events like the Covid-19 pandemic, reshape the way people travel.

Accessibility measures how people and activities between different locations can be reached through a transport network, an air transport one in that instance (Rendoni et al., 2013). The development of air transport networks has increased connectivity around the globe, with more city-pair and origin-destination connections, which result in the formation of a worldwide air transport network, with a shape similar to the one shown in Figure 4.1. The most interesting finding of this research is that despite their complexity, networks ranging from global air traffic to neural ones share very similar backbones made of essential nodes and links, similar to thick routes (very busy routes) within the international, regional or national air transport network. In addition, there are links that are important or not important to each node, not only within an air transport network but also in almost all network types, even within the human body.

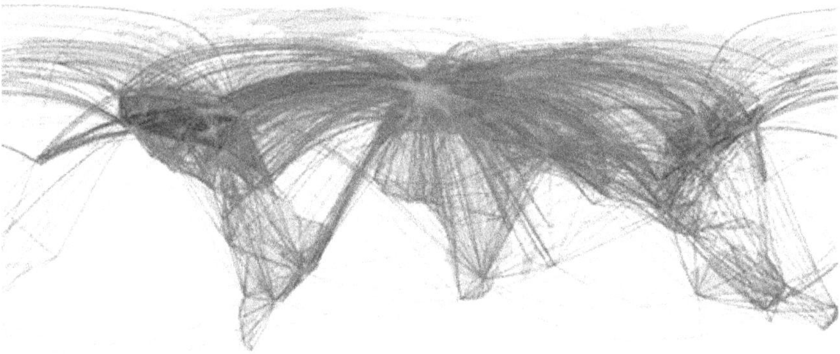

Figure 4.1: The worldwide air transport network. Source: Northwestern University, 2012

Within the air transport network, the decisions on establishing an important or not important node (connection) are not determined by the laws of nature or human body, but by passenger demand. Passenger demand though can be intrinsic or extrinsic, therefore airlines, airports and destinations try to generate demand by a number of incentives and concessions. Wittmer and Laesser (2009) argue that while on the one the hand businesses take into consideration air transport access when making location decisions, on the other hand air pollution and noise that derive from increased air transport movement lead to requests to limit their development. At the same time, there is a link between air travel frequency and the limitation of the number of connections (Wittmer & Laesser, 2009). This link reinstates the argument posed earlier that air connectivity generates demand since it improves accessibility locally, regionally and internationally, making it easier for passengers to travel as it offers choices of destinations and access to the airport is relatively easy, at least in terms of travelling time. The above statements dictate that air transport connectivity is very complex, therefore this chapter focuses only on traffic for leisure purposes and its link to tourism.

Air connectivity is vital for the accessibility of destinations whatever their size or popularity. There are destinations which are served by very large aircraft carrying hundreds of passengers and others that are served by small aircraft carrying a few dozens of passengers (PwC, 2017; IATA, 2020; ICAO, 2021). Despite aircraft size and passenger volume, accessibility and air connectivity are of paramount importance in terms of passengers' mobility and destination success, popularity and development. There are numerous factors that define air connectivity and these are not limited only to infrastructure limitations or supply and demand levels. International regulations, bilateral or multilateral agreements, geopolitical and equipment availability are some of the factors that affect air connectivity.

Traditionally, there are connections between some countries, however these are dynamic relations which are shaped over time depending on the current market and financial conditions. The aim of this chapter is to address most of these relationships, namely air connectivity both at international and regional levels, and analyse the relationship between air connectivity and leisure traffic and finally discuss the challenges and applications between them. The chapter is based primarily on air transport related literature with a tourism scope. It addresses how technological and equipment developments can reshape connectivity nodes and networks which in return will reshape air connectivity for leisure travel in the future. This is achieved by presenting a conceptual contribution based on current literature in the field of contemporary tourism geography, economic geography and tourism mobility.

International connectivity and regional connectivity

As discussed above, there are several methodologies used to measure connectivity and several factors affect connectivity. There have been numerous studies evaluating air transport networks and their connections, some of which focus on the network structure and performance whereas others focus on connectivity and centrality (Morrison & Winston, 1995; Allroggen et al., 2015; Burghouwt, 2017; Cheung et al., 2020). When referring to connectivity we tend to mean the degree of connection between airports or nodes as described in the previous section. There are two key factors that define connectivity between two airports (destinations), one being frequency of services and the other directness. The first is self-explanatory as it refers to the number of services (flights) available from point A to point B over a given period of time, usually daily or weekly. When referring to services between two points, it is vital to determine that there are two distinct types of connectivity, direct and indirect. Indirect connectivity within the air transport industry refers to the number of possible connections one can choose from to fly from one destination to another via one or more stops where the passenger changes planes (IATA, 2020). For example, a passenger can fly indirectly from Istanbul to New York by changing flights at another airport in Europe or can fly directly from Istanbul to New York.

From a passenger's perspective, the ability to travel directly by air from point A to point B in the shortest possible time is amongst the most important factors when choosing where to travel for leisure purposes, let alone business ones (IATA, 2020). These factors are even stronger when we refer to air connectivity for leisure purposes although there are scholars who argue for both sides. In essence, passengers who choose to fly for leisure purposes do consider the connectivity options when it comes to reaching their final destination and might choose to fly to a destination which is available by a direct flight from a nearby airport, instead of flying indirectly with a change of planes at another airport.

In the following two figures, the connectivity between and within major geographical areas is presented. They highlight the increased traffic in Asia and from Asia to traditional markets/geographical regions, namely Europe and North America. These trends and figures collapsed over 2020 as a result of the global pandemic, which is discussed in detail in the next section.

It has been argued by several scholars and studies that the introduction of low-cost carriers has played a crucial role in increasing connectivity, first within geographical regions and more recently, intra-regionally since the introduction of long-haul low-cost carriers.

What is missing from these pictures are the small regional airports. There are over 300 small and very small airports in Europe alone serving just over 8% of total European traffic according to 2011 figures (Redondi et al, 2013). Based on these findings, it can be argued that these airports are too small to form a major part in the networks and passenger flows. However, when examining some of these airports closely, it is evident that these play a vital role in the local society and economy, let alone the seasonal nature of traffic if some of them are located in tourism destinations. At the same time, the introduction of routes by low-cost carriers, not only in Europe, to underserved, uncongested and underused airports, even on a seasonal basis, has led to increased economic and social benefits both for the airports and the regions where they are located.

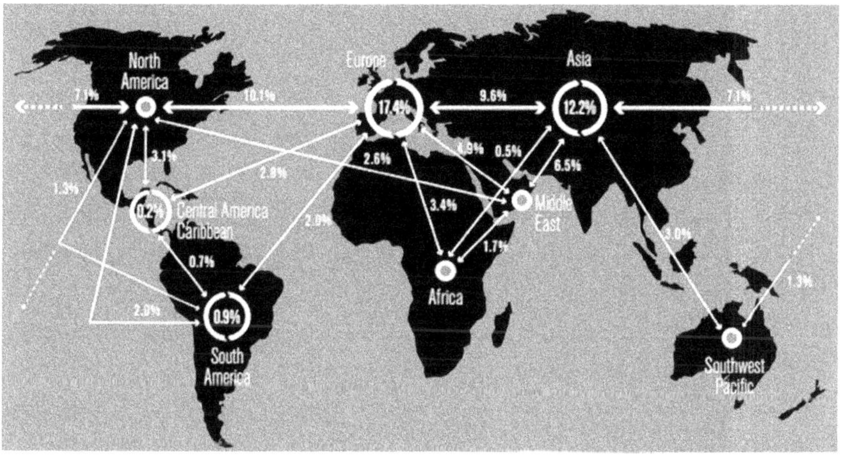

Figure 4.2: Percentage of international scheduled Revenue Passenger Kilometers in 2018. Source: IATA, World Air Transport Statistics, 2018

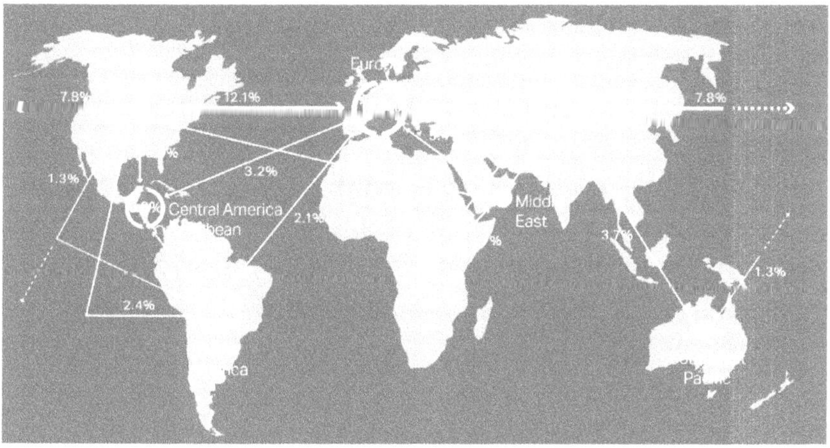

Figure 4.3: Percentage of international scheduled Revenue Passenger Kilometers in 2019. Source: IATA, World Air Transport Statistics, 2019

During the last quarter of 2019 there were some signs that the aviation industry and passenger demand would slow down which led several carriers to start discussing the withdrawal of widebody aircraft from routes that were not particularly profitable or did not demonstrate signs of strong demand. Several widebody aircraft had particularly high operating costs so airlines entered a phase of exploring options to either replace them or retire them completely. Demand for widebody aircraft had been slow even before the Covid-19 pandemic, with Airbus announcing the end of production of its A380 followed by a similar announcement by Boeing to cease production of the B747.

In 2020 the developments related to flying restrictions brought forward the decision on retiring ageing aircraft types which were uneconomical, like the B747 British Airways operated. At the same time, both these major widebody airframe manufacturers had been reducing production rates to meet the weakening demand by airlines (Aviation Week, 2021). In 2019, Airbus announced their plans to stop building its A380 superjumbo jet after the plane's biggest customer, Dubai-based Emirates Airline, cut its order by 39 planes. In addition, in a statement issued by Airbus in 2019 it was also stated that Airbus has *"no basis to sustain production, despite all our sales efforts with other airlines in recent years"* (Airbus, 2019). Emirates Airlines have been A380's largest customer and operator and their decision to switch their orders to smaller capacity aircraft, triggered a chain reaction by several competitors who decided to retire, earlier than expected, aged, cost inefficient, large widebody aircraft. Air France and Lufthansa have announced to retire their A380 fleets by 2023 whereas Asiana Airlines, Qatar Airways, Malaysia Airlines and Thai International Airways have revealed their plans to shrink or retire their A380 fleet, without providing a definite timescale (Forbes, 2020a).

In Spring 2020, it was reported that Airbus was facing financial difficulties, stating that they were unsure about the duration of this crisis and unprecedented decline in demand. One of the first measures that was implemented was to decrease aircraft production by over 50% compared to 2019 (Aviation Week, 2020).

Connectivity and leisure traffic

Increased connectivity results in numerous benefits to countries, regions and destinations as decreased travel cost and time stimulate demand, and low-cost airlines have been pioneers in that front (Burghouwt & Wit, 2015; Burghouwt, 2016; Burghouwt, 2017). At the same time, it is not until recently that scholars have started investigating the impact of air transport policy on tourism development (Forsyth, 2006; ComMark, 2006; Graham et al., 2008; Duval & Schiff, 2011; Dobruszkes & Mondou, 2013). There are around fifty very large airports in the world which account for over 34% of global capacity with the remaining being shared among thousands (Cheung et al., 2020). Leisure traffic tends to take place from high density population areas to less populated areas and the demand, especially at highly seasonal destinations, is stronger at one end of the airport pairs. This abnormality results in a positive skewed strength distribution with passenger flows concentrated on a few routes and the remaining having small numbers (Cheung et al., 2020). Given that the airline-airport relation is not vertical, unless there is mutual interest by both parties, the majority of airports where demand is weak or seasonal struggle to face competition, resulting in weaker or seasonal connectivity supplied primarily by limited-service carriers. The lack of continuous connectivity has adverse impacts not only on the airport but on the destination too, as continuous connectivity would make access to that destination faster and more cost effective.

Paros is a Greek island in the Central Aegean Sea, approximately 150km (93 miles) south east of Athens. The island has under 14,000 inhabitants, who can reach Piraeus on a three-to-five-hour sea journey, depending on the type of vessel and the number of ports it is calling at. Alternatively, they can arrive to Athens International Airport by scheduled flights in less than 40 minutes. The island's airport was built in 1982 with a runway under 600m (approx. 1970ft) (HCAA, 2019). This short runway did not allow international flights to use this airport and restricted the size of aircraft for domestic services to 20 seat capacity. A new airport was completed in 2016 with a runway of 1400m (approx. 4600ft) which resulted in a rapid growth of aircraft movements and passenger numbers, since jet aircraft could use this. Domestic passenger numbers almost tripled in three years and direct international services were launched in 2017.

Table 4.1: Aircraft and passenger movement change since the introduction of New Paros Airport

	Domestic		International	
Year	Aircraft Movements	Passengers	Aircraft Movements	Passengers
2016	2164	74,288	0	0
2017	3187	160,371	37	2602
2018	3652	200,507	52	4112

Source: Hellenic Civil Aviation Authority (HCAA), Statistical data of Paros airport for the period 1994 - 2018

Several regions and local authorities, in order to attract airlines and increase air connectivity, have invested in improving, upgrading or even building an airport, as it is anticipated that lack of or dated infrastructure is the main reason for not attracting enough leisure traffic. The reality is that infrastructure alone is not enough to generate enough connectivity or demand (Zhu et al., 2019). At the same time, several countries around the world are trying to stimulate traffic to those underserved airports or regions when there is no strong demand or commercial interest by airlines. Fageda et al. (2018) summarise that there are four main policy types introduced by governments to stimulate or sustain a degree of air connectivity between these regions, and these are route based, passenger based, airline based or airport based. There are countries like the United States of America which apply more than one policy type and at the same time there are countries in the EU which somehow have an approach that is not applied in every EU member state. Portugal and Spain, for instance, have offered discounts to residents when the majority of EU policies are either airport or route based. These policies offered residents discounts on air services, compared to non-residents. The main driving force for these policies has been social cohesion and promoting tourism to facilitate economic growth and development, which should result in increased social cohesion in return (Williams, 2010).

Ways forward to restore air connectivity

During the first wave of the COVID-19 pandemic, travel restrictions were imposed on several countries around the world and compulsory suspension of flights was introduced. This resulted in a severe loss of traffic and revenue both for airlines and destinations (IATA, 2020). The loss of connectivity due to the pandemic has been massive, the largest ever disruption in the modern history of tourism and air transport that resulted in a decline of over 70% compared to the previous year. UNWTO (2021b) estimates that the decline in international tourism arrivals was 74% compared to 2019.

Despite the sharp decline in international tourism arrivals and connectivity, domestic air traffic, as well as domestic tourism arrivals, performed significantly better with the decline in the region of 50% compared to the previous year, although there are significant differences across countries and regions. Domestic passenger traffic in China for instance was just 5% lower in November 2020 compared to the previous year (UNWTO, 2021b). According to ICAO (UNWTO, 2021b) domestic air capacity declined by 35% compared to 2019. Those significant fluctuations on performance, connectivity and tourism arrivals are due to the different policies, behaviours and perceptions of local residents who might have felt safer to travel domestically compared to internationally. At the same time, it must be said that international travel restrictions were implemented to a stricter and higher degree than domestic restrictions which allowed for residents of a country to travel within its borders.

The global pandemic is still unfolding in 2021 and although there are not as many restrictions as they were during the first phase, travelling globally is restricted to essential travel by nations and countries imposing strict restrictions on people who are willing to travel nationally as well as internationally. These actions have resulted in a sharp decline of demand and although services (flights) are not restricted, passenger demand is so low that several carriers around the world have suspended or reduced services to destinations.

However, there are essential services which are somehow supported by the government in order to maintain vital links between the periphery and core areas. The most known are the Essential Air Services in the United States and the Public Service Obligation (PSO) in Europe. These air routes are introduced when a government suggests that a minimum level of air connectivity and service is still required if the commercial interest is limited (Santana, 2009). These mandates differ from country to country but the general notion is that such air services are essential for social purposes to remote or low populated regions, where accessibility to major urban areas is crucial and limited to justify a financially sustainable commercial operation.

According to ACI (2021), the recovery in Asia Pacific to 2019 levels is expected to return in 2023 and it is predicted that recovery is going to follow a diverse pattern. A similar study by UNWTO (2021a) published in early 2021 suggests that recovery in the international tourism market is expected to commence in 2022. At the same time, there are significant differences in the responses of the respondents, based on their location with Europeans being more optimistic stating that in Q3 2021 recovery will be well on its way. On the other hand, residents from the Asia Pacific region are the more pessimistic since over 64% of them suggest that recovery will commence in 2022 followed by America where 58% of the respondents also suggest that recovery will commence in 2022.

One of the factors that affect passengers' and tourists' perception of safety during this crisis is how crowded a place can be (Li et al., 2020). Moreover, public transport restrictions in conjunction with local restrictions within a country might not make it easy for people to travel to airports further from their bases or their final destination. During the summer and autumn 2020 for instance, several countries had introduced lists, according to which arriving passengers after their trips abroad were liable to either self-isolation or not. Within countries, zones or tier systems were introduced, restricting or allowing residents and visitors to travel to or from these areas. Following these developments, several carriers have introduced direct international services from regional airports to popular destinations for 2021 in order to minimise those adverse impacts of travel restrictions.

Despite the promising messages from the public who are eager to travel again, the pandemic has worsened leading to stricter travel restrictions, including compulsory testing prior to departure, self-isolation and border closure similar to the first phase of the pandemic (Li et al., 2020). At the same time, spikes in cases within specific countries result to travel restrictions within the country having an impact on domestic tourism too. The public's confidence and anticipation were rising given that vaccination programmes had been introduced in several countries, however the pace of vaccination differed significantly from country to country. These developments have resulted in a prolonged period of uncertainty and continued weak signs of demand by those who want to make travel arrangements for the coming months (Karl et al., 2020). The general notion is that the beginning of recovery of the tourism sector is dependent on the pace of vaccination and the coordination amongst countries to ensure that the whole travel industry restarts.

Many airlines are still reducing frequencies and capacity to popular destinations when, other carriers announce new routes and services in light of generating demand and a stream of cash flow (Forbes, 2020b). The introduction of narrow body long haul aircraft like the A321 neo XLR (extra-long range) and reintroduction of B737 MAX with a range of 4700 nautical miles and 3800 respectively might be the answer to quick and swift recovery. Those aircraft given their smaller size and reduced operating costs might be the ones preferred by airlines to increase the fractured connectivity in the months and years to come. Twin engine aircraft were introduced by several carriers to connect long haul destinations outside major hub airports, like Manchester, Birmingham, Glasgow and Cologne to Dubai, Dallas, Philadelphia and so on. The introduction of smaller aircraft can be based on the knowledge and experience gained by airlines over the larger aircraft types which can lead to a steady and smooth recovery of the sector.

Air connectivity for leisure and tourism: challenges and applications

Air travel is the most popular choice of travel for leisure tourism (UNWTO, 2021b). In island destinations in particular, air connectivity is vital for their popularity and success, as air travel is significantly faster compared to other means of transport. Infrastructure on its own is not enough to create connectivity as the nearby destination served by the current airport has to have the potential to generate connectivity. The introduction of low-cost services changed air connectivity dramatically as these carriers were adventurous and started operating to remote airports which were adjacent or not that close to the established or known destination(s) (Zeigler et al, 2017; Joshan & Maertens, 2020; Ma et al., 2021). This approach has been successful by most low-cost carriers since there have been routes and airports that have sustained the pressure over the years and have established a pattern of connectivity. This has been achieved by the highly competitive pricing strategies which were implemented to generate demand and have been successful. Competitive pricing strategies are still in place today, not only by low-cost carriers, but these are the ones that tend to be more aggressive on pricing. In addition, low-cost carriers tend to be more adventurous when it comes to serving new destinations compared to other types of airlines.

This can be easily justified as a low-cost carrier, more often than not, does not carry passengers who are travelling on a packaged holiday. This approach means that the majority of passengers on board have made their own arrangements for their accommodation as well as access to/from the airport, without having arranged for a transfer service as part of a packaged holiday. At the same time, one could argue that low-cost carriers today offer packaged holidays but the majority of those offered are at destinations that are established within the low-cost carrier's network and generate high demand. In addition, several low-cost carriers might add a destination on their schedule only for a short period of time in the year just to meet demand which exists at their acceptable levels only during this specific period. There are numerous examples around the world where this approach materialises over the busy periods of these destinations, whether that being the skiing season or the summer holiday (Zeigler et al., 2017; Joshan & Maertens, 2020; Ma et al., 2021).

Non-continuous air connectivity is challenging for the destination for various reasons. To begin with, it is challenging for the airport infrastructure to become financially viable, as the fluctuating passenger and aircraft movement numbers do not allow these airports to generate enough income (Guimera et al., 2005; Cheung et al., 2020). There are hundreds of airports around Europe alone that serve less than 1 million passengers per year. These numbers are not healthy to allow for airport investment, hence the majority of these airports are state owned or municipality owned, and used as vehicles to promote the destination or region. These airports are financially supported by the authorities, local, regional or central government, as they are not commercially viable. At the same time, businesses supported by this visitor and tourist economy cannot totally rely on these air services since the passenger numbers and the duration of the service on an annual basis is not enough for their business hence, they rely on other transport types or markets to be financially viable (ACI, 2019). On the other hand, there are businesses which also operate on a seasonal basis and are not the main source of income for their owners. In that case, there is a tendency not to be interested in prolonging the service period of connectivity. Moreover, seasonal characteristics like snow do not allow for prolonging the serving period when everything at a destination is linked or concentrated in tourism products and services linked to snow. Investment and creativity are required to generate tourism products and services which are not directly linked to snow, and the most challenging part, to attract tourists outside this period or disassociate the destination with a specific type of holidays or activities. This is the most challenging as it is not just a matter of attracting tourists to the destination, but also of persuading the need for the local community and businesses to realise the need for change, embrace it and implement it.

Change is the only stable factor in the airline and tourism business. Both industries work hand in hand to provide services to the expanding and demanding tourism market. The business models both for the airline industry, the destinations, the accommodation and service providers are adapting to the demands and the needs of the market, although there are times that they precede them, as with the low-cost carriers. Those low-cost carriers were the ones who paved the way to a new way we fly and legacy carriers followed soon after. Their approach has been successful since more low-cost carriers are profitable compared to legacy ones. These accumulated profits allow them to try new products and services much easier and only time will tell whether they are successful or not.

Conclusion

Air connectivity for leisure destinations has been vital for their success especially as access is faster by air. This is more evident in island destinations as well as remote ones where surface access is challenging. It is highly likely that the technological improvements in aircraft technology, in line with the increasing trend, due to the Covid-19 pandemic, to travel to the final destination faster, avoiding connecting flights, will reshape air connectivity in the years to come. Airlines around the world are changing their fleet structure by withdrawing larger, cost inefficient aircraft, with ones which fly long distances and are smaller in size, compared to the widebody ones. Low-cost carriers have been leading the change towards increased air connectivity, followed by legacy carriers soon after. Operating costs directly linked to aircraft operations have been constantly decreasing allowing for more competitive pricing. At the same time, technological improvements have allowed aircraft to fly longer distances, introducing different sizes of them which allows all carrier types to adjust their fleet structure and serve a diverse range and type of destinations. This development has resulted in the introduction of several point-to-point services, which generated increased demand for leisure traffic. Since the only constant in the tourism and air transport industries is change, we have to see it unfolding in the years to come. The application of geography and mobility related literature underlines the theoretical contribution of this chapter based on air transport literature which expands and touches upon leisure tourism and mobility.

This chapter has been based on secondary data sources without any empirical findings, hence an empirical study at various spatial levels in the years to come will validate or not what has been discussed. It is more likely that regions will respond at different paces on the forthcoming air connectivity challenges that are emerging so this will be an interesting area to be studied. The chapter did not address the possible responses from airports or destinations to these potential changes that are likely to unfold in the years to come, hence this is an area that can be researched and explored by scholars in the fields of destination management, business and marketing.

Highlights

- ☐ Air connectivity is vital for the success of destinations, especially if they are islands.
- ☐ Infrastructure constraints do not always limit air connectivity, although improving them contributes to increased connectivity.
- ☐ Low-cost carriers contributed significantly to increased regional connectivity.
- ☐ New aircraft types have, are and will impact significantly on air connectivity across different levels.

Questions

1. Will air connectivity recover to the pre-COVID-19 period or will new patterns and connections emerge?
2. Will the point-to-point connections remain popular or will we experience a return to the hub and spoke network structure?
3. Will the introduction of long haul, single aisle aircraft with carrying capacity of up to 200 passengers create new air connectivity opportunities?
4. Will the introduction of these aircraft be the opportunity for low-cost airlines to operate on long haul routes with aircraft belonging in the same family (B737 MAX or A321)?

References

ACI Asia Pacific (2021). *Long Term Recovery of Airport Sector Contingent on Short Term Actions*. Press Release.

ACI Europe (2019). *Airport Industry Connectivity Report*.

Airbus, (2019). *Airbus and Emirates reach agreement on A380 fleet, sign new widebody orders*. https://www.airbus.com/newsroom/press-releases/en/2019/02/airbus-and-emirates-reach-agreement-on-a380-fleet--sign-new-widebody-orders.html.

Allroggen, F., Wittman, M., & Malina, R. (2015) How air transport connects the world – A new metric of air connectivity and its evolution between 1990 and 2012. *Transportation Research Part E*, 80, 184-201.

Aviation Week (2021). *Why The Widebody Market Will Remain Weak For Years*. 04 January.

Aviation Week (2020). 'Survival of Airbus' At Risk, *Faury Warns*, 27 April.

Burghouwt, G. (2016). Focus on benefits: Assessing consumer welfare impacts of aviation policy measures Airline responses, lumpy capacity and hub rationalization. *OECD International Transport Forum*. www.itf-oecd.org/sites/default/files/docs/asssessing-consumer-welfare-impacts-aviation-policy.pdf.

Burghouwt, G. (2017). Influencing air connectivity outcomes. *International Transport Forum* Discussion Paper.

Burghouwt, B., & Wit, W. (2015). On the mechanisms that can potentially influence connectivity outcomes in the UK (SEO Report No 2015-40). www.seo.nl/uploads/media/201540_On_the_mechanisms_that_can_potentially_influence_connectivity_outcomes_in_the_UK.pdf.

Cheung, T., Wong, C., & Zhang, A. (2020). The evolution of aviation network: Global airport connectivity index 2006-2016. *Transportation Research Part E*, 133, 101826.

Coles, T., & Hall, C.M. (2006). The geography of tourism is dead: Long live geographies of tourism and mobility. *Current Issues in Tourism* 9(4-5), 289-292.

Coles, T., Duval, D., & Hall, C.M. (2004). Tourism, mobility and global communities: New approaches to theorising tourism and tourist spaces In W. Theobold (Ed.) *Global Tourism*, 3rd ed. (p.p. 463-481). Oxford, UK: Heinemann.

ComMark. (2006). Clear skies over Southern Africa, The importance of air transport liberalisation for shared economic growth. Johannesburg: ComMark.

Dobruszkes, F., & Mondou, V. (2013). Aviation liberalization as a means to promote international tourism: The EU-Morocco case. *Journal of Air Transport Management*, 29, 23–34.

Duval, D. T., & Schiff, A. (2011). Effect of air services availability on international visitors to New Zealand. *Journal of Air Transport Management*, 17(3), 175–180.

Fageda, X., Suárez-Alemán, A., Serebrisky, T., & Fioravanti, R. (2018). Air connectivity in remote regions: A comprehensive review of existing transport policies worldwide. *Journal of Air Transport Management*, 66, 65-75.

Forsyth, P. (2006). Aviation and tourism. In L. Dwyer, & P. Forsyth (Eds.) *International Handbook on the Economics of Tourism* (p.p. 581–623). Cheltenham, UK: Edward Elgar.

Forbes, (2020a). *The last A380 just rolled off Airbus' production line. it may never fly a single passenger.* 25 September. https://www.forbes.com/sites/erictegler/2020/09/25/the-last-a380-superjumbo-just-rolled-off-airbus-production-line-will-it-ever-fly-passengers/.

Forbes (2020b). *The Pandemic Is Forcing Big Airlines To Make Profound, Long-Lasting Decisions About Their Fleets.* 09 November. https://www.forbes.com/sites/danielreed/2020/11/09/the-pandemic-is-forcing-big-airlines-to-make-profound-long-lasting-decisions-about-their-fleets/.

Graham, A., Papatheodorou, A., & Forsyth, P. (2008). *Aviation and Tourism: Implications for leisure travel.* Aldershot: Ashgate.

Guimera, R., Mossa, S., Turtschi, A., & Amaral, L.A.N., (2005). *The Worldwide Air Transportation Network: anomalous centrality, community structure, and cities global roles.* PNAS 102, 7794-7799.

Hall, C.M., Amore, A., & Arvanitis, P. (2019). The globalising force of human mobilities, in D. Timothy (ed), *Handbook of Globalisation and Tourism* (p.p. 55-65). Massachusetts: Edward Elgar.

Hellenic Civil Aviation Authority (HCAA) (2019). *Statistical data of Paros airport for the period* 1994 – 2018. http://www.ypa.gr/en/our-airports/kratikos-aerolimenas-paroy-kapa.

IATA (2020). *Air Connectivity: Measuring the connections that drive economic growth.* IATA, Montreal.

IATA (2019). *World Air Transport Statistics,* Plus Edition 2019. Montreal - Geneva.

IATA (2018). *World Air Transport Statistics,* Plus Edition 2018. Montreal - Geneva.

ICAO – International Civil Aviation Organisation (2021). *Air Transport Accessibility*. Montreal.

Joshan, S., & Maertens, S. (2020). Low cost carriers in the Middle East and North Africa (MENA) region: Emergence and barriers to development. *Journal of Transport Geography*, 87. doi: 10.1016/j.jtrangeo.2020.102799.

Karl, M., Bauer, A., Ritchie, W.B., & Passauer, M. (2020). The impact of travel constraints on travel decision-making: A comparative approach of travel frequencies and intended travel participation. *Journal of Destination Marketing and Management*, 18. Doi: 10.1016/j.jdmm.2020.100471.

Li, Z. Zhang, S., Liu, X., Kozak, M., & Wen, J. (2020). Seeing the invisible hand: Underlying effects of COVID-19 on tourist's behavioral patterns. *Journal of Destination Marketing and Management*, 18. Doi: 10.1016/j.jdmm.2020.100502.

Ma, W., Zhang, A., Zhang, Y., & Xu, S. (2021). The growing influence of low-cost carriers in Northeast Asia and its implications for a regional single aviation market. *Journal of Air Transport Management*: 91. Doi:10.1016/j.airtraman.2020.101994.

Morrison, S.A., & Winston, C., (1995). *The Evolution of the Airline Industry*. Brookings, Washington DC.

Northwestern University, (2012). *Researchers Show Complex Networks Have Same Skeletons*. Northwestern University. https://www.mccormick.northwestern.edu/news/articles/2012/06/article_brockmann-networks-skeletons.html.

PricewaterHouseCoopers (2017). *Connectivity and Growth*.

Redondi, R., Malighetti, P., & Paleari, S. (2013). European connectivity: the role played by small airports. *Journal of Transport Geography*, 29, 86-94. doi.org/10.1016/j.jtrangeo.2013.01.010.

Santana, I, (2009). Do public service obligations hamper the cost competitiveness of regional airlines? *Journal of Air Transport Management*, 15, 344-349.

UNWTO (2021a). *Worst year in tourism history with 1 billion fewer international arrivals*. https://www.unwto.org/news/2020-worst-year-in-tourism-history-with-1-billion-fewer-international-arrivals?fbclid=IwAR38UR-Wy1H5xFOrvRh88_laGdgQFYk28N9364icmokaDhDpvW-Ycwb7miI

UNWTO (2021b). *International Tourism Highlights*.

UNWTO (2019a). *International Tourism Highlights*.

UNWTO (2019b). *International Tourist Arrivals to reach 1.4 billion two years ahead of forecasts*. https://www.unwto.org/global/press-release/2019-01-21/international-tourist-arrivals-reach-14-billion-two-years-ahead-forecasts.

Williams, G. (2010). European experience with direct subsidization of air services. *Public Money Management*. 25, 155-161.

Wittmer, A., & Laesser, C. (2009). Customer issues related to limited air connectivity. *Journal of Air Transport Management*, 15, 23-25.

Zeigler, P., Pagliari, R., Suau-Sanchez, P., Malighetti, P., & Redando, R. (2017). Low-cost carrier entry at small European airports: Low-cost carrier effects on network connectivity and self-transfer potential. *Journal of Transport Geography*, 60, 68-79.

Zhu, Z., Zhang, A., Zhang, Y., Huang, Z., & Xu, S. (2019). Measuring air connectivity between China and Australia. *Journal of Transport Geography*, 74, 359-370.

5 Tacit hierarchising in online communities of hillwalkers

David Brown and Sharon Wilson

Abstract

This research explores how Munro-baggers – hillwalkers aiming to climb all 282 Scottish mountains over 3,000ft – hierarchise themselves and others as serious leisure participants. This increasingly popular hobby contributes to Scotland's economy and profile, but its sparse literature insufficiently analyses the influence of Stebbins' Serious Leisure Perspective (SLP), the recent reappraisal of Serious Leisure or the influence of online communities. Therefore, we critically revisit the SLP to re-evaluate Munro-bagging. Through phenomenological interviews, we explore how Munro-baggers hierarchise each other, tacitly and otherwise, offline and online, through their activities' perceived characteristics. Ambiguities and overlaps are explored and the interplay of contexts analysed. We identify factors influencing Munro-baggers' perceptions of seriousness amongst fellow hobbyists, taxonomising participants by their perceived characteristics of seriousness. Findings suggest that they draw upon quantitative and qualitative judgments of hobby-relevant activities and qualitative judgments of certain *ad hominem* characteristics. The expansion of the pastime beyond its temporospatial boundaries into online spaces is found to influence the extent to which actors categorise or hierarchise each other and the characteristics used to do so.

Keywords:

Munro-bagging, serious leisure, hillwalking, adventure tourism, mountain tourism, ethnography, netnography

Introduction

Hillwalkers, mountaineers and climbers, who toil in pursuit of 'serious leisure' (Stebbins, 1982), increasingly use the internet to immerse themselves remotely in leisure communities which would otherwise be denied by temporospatial boundaries. In this chapter, we explore a theoretical lacuna around digital mobilities as we link debates surrounding mobility studies and media communications (e.g. Hollett et al., 2017) with those discussing ephemeral ways of moving and being with social media (Møller & Robards, 2019). Exploring how physical work rubs up against the virtual, we examine online socialisation in the context of Munro-baggers, and how digital context influences their perceptions of self and each other – something neglected within extant theory. We discuss their apparently hierarchical, tacit or formal classifications and their influence upon Munro-baggers' experiences.

Munro-bagging is an expanding socio-cultural phenomenon largely neglected within academic literature. A form of serious leisure (Stebbins, 1982; Stebbins, 2007; Elkington & Stebbins, 2014), it entails systematically climbing the 282 Scottish mountains exceeding 3,000 feet (914 metres) high. Over 6,000 hillwalkers have 'bagged' every summit (Scottish Mountaineering Club, 2018), 'completing' a Munro 'round' and graduating from Munro-bagger to Munroist. Several times this number are partway through (ibid). The pastime is becoming more socialised through hillwalkers' websites, enabling peer-to-peer interactivity and user-generated content. This potentially broadens Munro-bagging, breaking its location-specific, temporospatial boundaries and providing additional contexts for participant interactions, construction of self, and community formation. This research explores how Munro-baggers gauge seriousness within online communities, and connects these insights to discourses as an evolution of new directions in tourism theory and practice. Taking a pursuit which has unfolded informally online, we demonstrate empirically the complex ways in which serious leisure tourists find new meanings in space and place, expressing primal competitiveness and social connection in a virtual space. Rather than simply suggesting that online opportunities are progressive and sustainable in the new epoch of travel developments, we also acknowledge that they are confounding and fascinating, as tourists are motivated towards 'being human' in a hydridic, computer-generated time-space configuration. We propose that a dystopia is also potentially developing, as participants elicit old behaviours by socially positioning themselves in an alternative playground, using technologies which may generate inequalities, as social orderings pervade beyond the leisure-scape.

Review of the theory

Extant classifications of serious leisure participants

As a progressive debate, we note several authors who have provided models of serious leisure stratification from its inception with Stebbins (2007) and the adoption of Unruh's (1980) classification of participants in social worlds – 'strangers', 'tourists', 'regulars' and 'insiders'. Building on recent conversations and acknowledging increases in online socialisation, we return to Stebbins (2017), who finds leisure behaviour increasingly complex and multi-dimensional. Lee and Ewert (2019) have reinvigorated debates around the motivations of serious leisure participants, whilst Heidari et al. (2019) have investigated leisure behaviour as a structural model of serious leisure and place attachment. Munro-bagging provides potential as a trope to understand tourism practices more broadly. Unruh's (1980) 'strangers' have little apparent understanding or appreciation of the pastime but may gain peripheral, unintended and uninterested experiences, often serendipitously. Therefore, this study explores these illusive practices where 'tourists' sporadically undertake the pastime, whilst 'regulars' make frequent Munro-bagging trips, and 'insiders' have sufficiently extensive skills, experience and networks to influence peers and access special resources – an interesting travel phenomena when extrapolated into online spaces.

Elkington and Stebbins' (2014) classification of serious leisure adherents as 'neophytes', 'participants', 'moderate devotees' and 'core devotees' differs from Unruh's (1980) taxonomy, focusing less on attitudes, more on 'career' progression. Brown's (2007) 'occasional', 'recreational', 'wannabe', 'competitive' and 'hardcore' categories reflect actors' assumed social positions, but their generic nature ignores the processes which Munro-baggers undertake, and actions observed online. Moreover, Stebbins' (2007) classifications largely overlook how leisure dynamics change over time (Veal, 2017) as actors develop.

Munro-bagging 'seriousness' and leisure

Munro-bagging requires knowledge, planning, perseverance, fitness, self-sufficiency, stamina, navigation skills, determination and nerve. Considering Stebbins' (2007) serious leisure participants (SLP) categories of 'serious', 'casual' and 'project-based', Munro-bagging is 'serious' – a substantial, fulfilling pursuit requiring specialised skills and knowledge. Neither purely hedonic nor 'casual' leisure, its adherents make short-term sacrifices for long-term goals, deferring gratification – the breathlessness and fatigue of climbing to gain the summit, and the nondescript Munro to 'tick' an objective. The 'casual' leisure attributes of being short-lived, immediate and free of skill are inapplicable. Nor is Munro-bagging 'project-based', being both fulfilling and longitudinal. Rather, the six qualities of 'serious leisure' – perseverance, career progression, durable benefits, unique ethos, identification with the activity – and the need for effort, training, knowledge and skills (Elkington & Stebbins, 2014), are all present. 5

The SLP

The SLP (Stebbins, 1982) represented a generational advance in understanding leisure (Dilley & Scraton, 2010; Gallant et al., 2013), reconceptualising it (Shen & Yarnal, 2010). Stebbins (2012) described four leisure categories distinguishing serious leisure: systematic yet unpaid pursuit requiring special skills, knowledge and experience; devotee work, which is self-enhancing and feels like leisure but contributes to a participant's livelihood; casual leisure as hedonic, short-lived but intrinsically rewarding work requiring little or no training or skill; and project-based leisure which, whilst short-term, may require planning, effort or skill, or be reasonably demanding. He (ibid) allocated serious leisure six qualities: the need for perseverance; the need for skill, knowledge or training; its potential to become a career; its unique ethos; the sense of identity derived from it; and the durable benefits enjoyed by participants, such as self-fulfilment, belonging, self-expression and fun. Though comprehensive, Veal (2017) found it neglected the significance of other areas of life, such as work and family, in terms of self-actualisation, which shapes 'leisure lifestyle'.

Serious leisure may underpin a social world of actors, organisations, practices and events binding together participants (Unruh, 1980) and providing meaning, solidarity and self-identity (Rojek, 2000). Such communities are distinctive, recognisable to members, and contain subcultures with norms, belief systems, values, styles, guiding principles, moral standpoints, and standards by which members are judged (Unruh, 1980). Serious leisure may therefore be a conduit to nurturing community (Gallant et al., 2013), bestowing social benefits (Brown, 2007) such as cultural nourishment, entertainment and relaxation (Dumazedier, 1974). Elkington and Stebbins (2014) extended the SLP by adding geography, arguing that it is usually space-specific. However, the bourgeoning nature of online community participation amongst leisure enthusiasts has challenged the geographical specificity of even those pursuits which appear located in one place, such as Munro-bagging. Stebbins (2006) suggested that the contemplation and reflexivity permeates serious leisure – perhaps explaining why Munro-baggers often elongate the experience pre- and post-activity and justifying linking involvement and commitment in serious leisure theory (e.g. Iwasaki & Havitz, 2004).

Serious leisure and identity

Taylor and Kay (2015) partially anticipated Veal's (2017) request for greater consideration of social identity by exploring how engagement underpins participants' self-identities and the meaningfulness of their occupations. Veal's (2017) focus on pursuit specialisation and its effects was partially foreshadowed by earlier calls to link the SLP and specialisation (e.g. Lyu & Oh, 2015) – although these remain largely unanswered. More robust challenges to the SLP emanated within feminist studies, and especially the accusation that 'serious' is masculinised, overlooking the importance to females of talk, humour and friendships in addressing normative gender discourses and constructing subjective identities (Green, 1998). Shaw (2001) asserted that, to overcome gendered constraints, inequalities and oppression, women use 'leisure as resistance', so individuals' empowerment through the reconstruction of entrenched, gendered ideologies should underpin the serious leisure concept. Wearing's (1998) shift to poststructuralist analysis enabled deeper exploration of diversity, intersectionality and difference across multiple sites of power, especially through 'leisure as space', enabling women 'enlargement of the self' (p.188). This has enabled leisure's role to be understood within the contexts of women's lives, social relationships, physicalities, attitudes, gender identities, sexual relationships and work (Dilly & Scraton, 2010), rather than these factors being considered discrete and separable.

Research design, method and data collection

The first author has been a Munro-bagger since 2001 and has informally frequented Munro-bagging websites since around 2012. From 2019, formal, empirical, 'insider' research commenced to observe 'digital tribes' (Cova et al., 2007) and the online socialisation of identity-construction on hillwalkers' websites. A preliminary netnographic approach addressed Elkington and Stebbins' (2014) lack of focus on communities and cliques highlighted by Veal (2017). Rather than analysing specific interactions and adopting a purely netnographic approach, broader a priori themes were explored in ethno-phenomenological interviews with Munro-baggers. Questions were minimal and open-ended, and participants led the discussions, in which 'seriousness', 'leisureliness' and hierarchised perceptions of Munro-bagging activity emerged (though not expressed in those terms). 5

Although the research makes no claim for generalisability, it used a broadly representative sample of Munro-baggers. Few data exist on Munro-baggers but most participants are from Scotland, and a large minority from England and Wales (SMC, 2018). Although women comprise around 40% of Munro-bagging website users, it is nearer 25% on the hill. Most Munroists take over two decades to complete their round, although some take a lifetime and others complete quickly, a tiny minority undertaking continuous rounds (SMC, 2018). All participants were consciously Munro-bagging. Convenience sampling was used, with the author recruiting hillwalking friends, inviting participants through social media, and contacting participants recommended by the first and second wave of interviewees. The initial participants recruited a further 16 participants. Whilst two were Munroists (had summited all the Munros), several were nearing 'compleation', most had climbed 10-30 Munros, and others were casually climbing Munros but making few sacrifices and did not expect to 'compleat'. The 28 participants were split exactly by gender and location between Scotland and Northern England. After 33 interviews, five participants provided insights in 20- to 50-minute follow-up interviews, before saturation was acknowledged in accordance with Guest et al.'s (2006) criteria. Interviews were conducted in neutral venues such as hotel lobbies, recorded digitally, and a research journal maintained of non-verbal communication, impromptu thoughts, emotions and observations. Transcripts were analysed by deep reading and Template Analysis to identify themes and sub-themes (King & Horrocks, 2010).

Findings

Factors considered in Munro-baggers' evaluations of themselves and others

Interview analysis presented 14 recurring factors which Munro-baggers consider when evaluating involvement. The first three are quantitative and address the objective of Munro-bagging. The first, most common, is the number of Munro ascents achieved – usually the number of *different* Munros summitted, as this is more congruent with Munro-bagging than repeating the same ascents, and therefore more 'prestigious'. Interviewees admired people who had '*amassed an impressive number*' or who – as new Munro-bagger, Janice described – were '*closing in on the magic 282*'. The second factor is trip frequency, although this allocated less prestige, as someone may have '*caught the bug pretty recently*' or adopt the pastime as '*their latest obsession*' whilst '*cherry-picking the easy Munros and ignoring the tougher ones*', ultimately neglecting the central objective of Munro-bagging. The third factor, length of experience, was considered a reliable indicator of dedication and attachment to the pastime, although Bobby, a Munroist, noted that '*some tough nuts start, finish in a couple of years, and disappear again*'. These first three factors are attributable to the participant's level of immersion in the hobby, with cross-sectional immersion (i.e. intense, possibly fickle, bursts of activity acknowledged in phrases like '*summits this year*') valued less than the longitudinal immersion of '*number of years' experience*' or '*total Munros bagged*'. As such they are progressive, allowing walkers to accumulate status.

The second group of factors comprise five qualitative criteria addressing Munro-bagging activity. The first is the inclusion of solo ascents. Most interviewees felt that solo walking demands more self-sufficiency, determination and '*bravery*' in coping with navigation, fatigue, injury and loneliness. Some, like Janice, said they were '*trying to pluck up courage to do a solo hike*' and that solo hillwalkers are '*gnarly, hardy beasts*'. The second factor was the inclusion of 'tops' within Munro-bagging rounds. 'Tops' are Munros' subsidiary peaks above 3,000 feet. Many require arduous detours. A third factor is the inclusion of foul weather or winter ascents. Scottish winter mountaineering requires additional skills (e.g. ice axe and crampon usage, rope work and avalanche risk assessment). Many summer hillwalkers consider winter mountaineering '*out of my league*' and '*strictly for experts*', whilst others pledge to '*get myself trained up when I have more confidence'* at outdoor education centres. The fourth factor is inclusion of multi-day expeditions or classic routes. For example, casual Munro-bagger, Amanda, observed that *'more serious hikers probably wouldn't do Ben Nevis via the Tourist Track with the day-trippers. They'd probably do the Carn Mor Dearg route or one of the gullies on the North Face.'* Such an approach was deemed *'purist'* and *'less mercenary'*, as ease of summitting was deprioritised below aesthetic appeal and physical challenge. The final factor in this group is the inclusion of complementary activities. Some, like bothying (i.e. sleeping in mountain huts), wild-camping, bivouacking or mountain biking, may be needed to reach remote summits, whilst others, like kayaking, may embellish the pastime, providing immersive experiences of nature. Participants doing this are often perceived as *'proper outdoors types'* and *'all-rounders'*. For all five factors in this group, observers judge seriousness qualitatively based upon style of Munro-bagging, rather than on numerical values.

The final group comprises six factors which are qualitative but *ad hominem* – participant-focused, not activity-focused. The first is the walker's overseas experience. Undertaking foreign climbs (e.g. Mont Blanc) is considered positive differentiation. One interviewee called it *'finishing school'* and *'going above and beyond'*, whilst another described how returning to the Munros from the Alps was *'a step down'*. The second factor concerns self-presentation. Although several interviewees criticised uncommitted walkers who possess *'all the gear but no idea'* (a common insult, possibly of army origin), most admitted to making assumptions based upon clothing and accessories. Nicola, halfway through her round, recalled an encounter with *'this old mountain goat in a Paramo Alta [expensive weatherproof jacket] which obviously cost him a fortune but was old and battered. You could just tell from his jacket he was a bit of a Jedi [i.e. – a time-served master]'*. Having the 'right' gear attracts admiration when worn for functionality and without ostentation. Arthur, an ex-soldier, noted that social media members *'seem to treat the old beardy guys with more reverence than some of them deserve. Some of the younger females who post summit selfies where they're wearing make-up get more gentle encouragement, like they're new to Munro-bagging or need a gee-up. But looking at their profiles, some of them have done more than the beardies.'* This reinforces that the androcentric, masculinised interpretations of serious leisure discussed in feminist studies (e.g. Heddon & Turner, 2012) influence perceptions of men's and women's seriousness. The third factor in this group – display of knowledge – is more meritocratic. Munro-baggers are perceived more favourably by offering advice on routes, mountain topography and geography, Highland history and culture, logistics and gear. Less experienced interviewees described such people as *'a mine of useful advice'*, *'a guardian angel in the chatroom'* and *'a mentor to new starters'*, whilst similarly experienced others sought them out as likeminded people.

The fourth qualitative but participant-related factor is membership of clubs, societies, chatrooms, committees or other social entities. One interviewee spoke admiringly of a walker they had met in a youth hostel who had been a member of the Creag Dhu Club, a seminal group of working-class Glaswegian trailblazers. The fifth factor focuses on the participant's social orientation – their ability to entertain, inform and amuse through anecdotes, interaction quality and community-mindedness. Several interviewees recalled admiring someone who *'was funny, empathetic and supportive'*, who *'loved introducing new people to the hills and sharing their passion'* or who was simply *'great value for money'*. Usually, their positive social attributes were understood in a Munro-bagging context, although the same traits were probably prevalent elsewhere in their lives. The final factor is display of an appropriate ethos. Most interviewees found walkers more attractive companions and accomplished participants if they displayed strong awareness of issues relating to mountain safety, conservation and nature, although there was an aversion to *'holier-than-thou types who preach about everything as if everyone else is some sort of yobbo [i.e. – miscreant]'*. Two interviewees felt that Scottish walkers are more readily respected than English ones due to Munro-bagging being a specifically Scottish pastime with an ethos distinct from walking in English areas like the Lake District, and differences in terrain, socio-cultural history, and land access laws.

As we have seen, when evaluating seriousness, Munro-baggers appear to consider three categories of factors. The first is quantitative and concerned with the central aim of the hobby, in which all walkers gain status as their 'career' progresses. The second is qualitative and concerned with the characteristics of trips undertaken, and some walkers may gain status by adopting practices optional to Munro-bagging. The third is also qualitative but is concerned with *ad hominem* characteristics rather than Munro-bagging activity and confers additional status only to those who make certain choices which are not prerequisites of Munro-bagging. These categories are summarised below.

Table 5.1: Categories and factors considered in Munro-baggers' evaluations of themselves and others

Quantitative, activity-focused factors
- ☐ Total Munro summits bagged
- ☐ Frequency of Munro-bagging trips
- ☐ Length of Munro-bagging career

Qualitative, activity-focused factors
- ☐ Inclusion of solo summits
- ☐ Inclusion of the Munro 'tops'
- ☐ Inclusion of Munros summited in winter or bad weather conditions
- ☐ Inclusion of multi-day or 'classic' routes
- ☐ Incorporation into trips of additional, related pastimes

Qualitative, ad hominem factors
- ☐ Foreign mountain experience
- ☐ Self-presentation / appearance
- ☐ Display of knowledge and expertise
- ☐ Membership of relevant clubs and societies
- ☐ Social orientation
- ☐ Understanding of ethos

Emerging categories of Munro-baggers by their characteristics

Having used the factors above to determine seriousness amongst peers, Munro-baggers appear to place each other into loose categories, with the most committed and accomplished members at the top and three further categories beneath – several interviewees specifically referred unprompted to a *'hierarchy'*, *'pecking order'* or *'climbing the ladder'*, suggesting that progression between categories is desirable. Whilst boundaries are indistinct because members of neighbouring categories may share several characteristics, four clear identities emerged from the data, here proposed as *'hardcore'*, *'committed'*, *'moderately committed'* and *'inwanderers'*.

'Hardcore' Munro-baggers are the most engaged, committed and skilled. Whilst the word *'hardcore'* was used by over half of interviewees, other terms included *'Jedi'*, *'demi-gods'*, *'mountain goats'* and *'old men of the mountains'*. These humorously irreverent descriptions emphasise masculinity and maturity, but were used by interviewees regardless of age and gender. Although 'hardcore' members may be new to Munro-bagging (e.g. if their previous mountaineering experience was acquired on difficult foreign terrain), they are perceived to have bagged many Munros. They are *'those gnarly types who take themselves off for days on end…climbing ridges called 'Suicide Precipice' or 'the Widow Maker'*. Whilst darkly humorous, this portrait is telling (and not only because climbs called 'Widow Maker' do exist). 'Hardcore' participants are perceived as physically formidable purists pursuing technical routes – highly self-sufficient, experienced practitioners who summon resourcefulness and perseverance over demanding, multi-day expeditions. Another interviewee mentioned that they '*know where all the secret howffs and dosses [both crude open shelters] are...always have a key to the locked climbers' huts*'. 'Hardcore' Munro-baggers, then, are insiders (Unruh, 1980; Stebbins, 2007) with exclusive resources and expert knowledge of their pastime, the environment and mountain ethos.

The second category is 'committed' Munro-baggers, who are hardier and more determined than the average walker, but not extreme enough to be 'hardcore'. They may be spotted '*trudging down off the hills in all weathers*' and may '*spend a small fortune every month driving up and down the A82*' to pursue their hobby. In other words, they are relatively undeterred and likely to be focused on the objective of climbing all 282 summits. As Michelle, a Yorkshire-based mid-career Munro-bagger, commented wryly, '*they've got their eyes on the prize*'. Others noted that '*they tend to have decent performance gear...wicking inner garments, ultra-lightweight waterproof shells...a lot of them seem to have little campervans*'. The 'committed' are perceived to invest heavily in their hobby and embed it within a broader lifestyle – connoisseurs rather than purists, as '*you'll see them taking classic routes like Ptarmigan Ridge on Ben Lomond, maybe even in winter conditions, but not really any hanging-on-by-your-fingernails' stuff*'. They are also portrayed as more social than 'hardcores' and their trips are carefully planned and less flexible – '*you meet quite a few of them in small groups...often middle-aged folks from further afield...always with a definite plan of where they want to go and how long it's going to take to get to point A and B*'.

The third category is the 'moderately committed'. Most interviewees considered this the most populous band. Although its members wish to climb all the Munros, '*most know they might never complete them all*', perhaps because '*they can only do a handful each year*'. The constraints of family obligations, distance and time are recognised. One interviewee commented that '*you might get these people doing the odd winter walk up one of the easier Munros, but nothing too technical and not in bad weather. You'd probably see them on more popular hills, togged up and looking the part but going up the main paths*'. These infrequent, 'fair-weather' Munro-baggers have modest technical prowess and little inclination to leave their comfort zones, fitting their walking around other activities and people, but displaying reasonable fitness and investing in appropriate gear.

The final category is the 'inwanderers'. Often newbies (Chayko, 2008) or neophytes (Elkington & Stebbins, 2014), they climb infrequently, having only loose plans. As one interviewee commented, *'they're a bit like the charity walkers you see on Beinn a' Ghlo – cheap new boots with no dubbin [waterproof grease] on them, plodding up cart tracks, no map, someone else telling them where to point themselves, no idea what they're climbing'*. Whilst possibly dismissive and elitist, this portrait describes an infrequent, uncommitted visitor with no specialist knowledge or expertise, little experience, low self-sufficiency, and little sense of adventure or mountain ethos. They closely resemble the unacculturated 'strangers' of Unruh (1980) and Stebbins' (2007) or Brown's (2007) 'occasional' participants. They may climb a specific mountain for its Munro status, Munro-bagging in the loosest sense, but have no intention to climb all 282 summits. Indeed, they may be unaware of a mountain's Munro status or ignorant of the concept. Therefore, the obscure word, 'inwanderer' (not to be mistaken for the German 'Einwanderer', meaning 'immigrant') is appropriate, being sufficiently broad to encompass new participants who wish to increase their engagement and those who have drifted uninterestedly into the pastime without expectations. The categories are summarised in Table 5.2:

Table 5.2: Categories of Munro-baggers by their characteristics

Hardcore	Committed	Moderately Committed	Inwanderers
Many summits bagged Very frequent trips Multi-day expeditions Self-sufficient Purist / 'classic' routes Physically demanding Exclusive / prohibitive participation Foreign / ultra-experience Athletic physique Appropriate gear Length of experience Technical skills Perseverance and endurance Sharing of experience Extensive knowledge Related mountain pastimes	Frequent visits Some winter and foul weather summits Good knowledge 'Classic' routes Length of experience Good gear Fit physique Some multi-day trips	Infrequent trips Fair weather walking Little winter experience Superficial knowledge Acceptable gear Healthy and fit Few / no technical skills 'Tourist' ascent routes Walking secondary to other holiday activities	Very infrequent trips Little / no experience No / few Munros bagged Reliant on others Easiest routes on 'honeypot' Munros Fair weather, summer walking Poor knowledge No technical skills Little appreciation of mountain ethos Not necessarily fit and healthy

The 'moderately committed' category, similar to Elkington and Stebbins' (2014) 'moderate devotees', is the second rung up of four, rather than their second rung down of four, as Munro-bagging is inherently committing. Therefore, here the median point is between 'committed' and 'moderately committed', whereas for Elkington and Stebbins (ibid) it is between 'participants' and 'moderate devotees'.

The hierarchising nature of Munro-bagger categories

Participants considered their perceptions of other Munro-baggers, and others' perceptions of them. Michelle, a 'compleatist', noted that participants could gain 'prestige indicators' [our phrase] as their hillwalking careers unfolded, but that this was neither important nor practicable for everyone: *'some people simply can't get here often enough or are happy just to potter along at their own pace. Night walking and wild camping aren't for everyone.'* This indicates that many participants do not aspire to more seriousness. Continuing, she commented that *'lots of people manage to bag all the Munros without going too far out their comfort zone – no wild camping, no blizzards'*. Therefore, one may fulfil the core aim of Munro-bagging with limited seriousness. Indeed, someone with 100 summits completed may be considered more 'serious' by peers than a Munroist if their methods were deemed more 'hardcore', reinforcing the importance of qualitative judgments and the fluidity of categorisations.

Asked about the recognition of categories, Geoff, an English walker with 18 summits bagged over ten years, noted that *'it's not a formal thing. There aren't specific labels, and people don't really think about it, but there is a pecking order, I suppose, and especially online'*. Like most other participants, Geoff believed that hierarchical categories exist, but perhaps unconsciously – tacit hierarchies of seriousness. Most interviewees considered categorisation an inevitable and innocuous aspect of human nature. As Josie, a Munro-bagging newbie from near Glasgow, commented, *'in any social situation, the first thing anyone does is have a sniff around the others to decide if they're their kind of people. You check out how they talk and dress... ...Oh God, when I say it like that, it's really depressing and shallow, isn't it?'* Josie appears correct in assuming that these judgments are heuristics enabling Munro-baggers to find peers whose experiences and aspirations match their own – facilitators of homophilous relationships. However, a minority of interviewees were irked by the phenomenon – for example,

Munroist Chris complained *'it's a bit pathetic really. When you finally get some freedom, it's a bit sad that people might still want to be top dog or to compete with other people. But I suppose counting Munros instils that attitude.'* Here, providing a wider context of belief systems and expectations, Chris appears half right. There is a hierarchised structure, but it appears more dependent on qualitative judgments about others' practices than the quantification of outcomes – Munro-bagging is product-driven, but participants make process-driven judgments. Several participants noted that the characteristics by which Munro-baggers judge each other are presented more plainly on websites than 'on the hill'. As Ian, midway through his Munro round, commented, *'it can feel like a dating site where all your vital statistics are laid out for the scrutiny of strangers – 'Munros climbed', 'years a member', 'number of comments', 'number of followers' – it can be a bit in-your-face'.*

By considering how these tourists organise themselves and interact online, we can extract implications for tourism, the travel industry, and of a social practice which denotes further opportunities within the serious leisure context. Direction changes, whilst uncertain, are revealed as competitive practices unleashed online, expressing inner tensions around social posturing. We have sought to contextualise these playful yet gently combative behaviours, towards new theoretical insights which could feed managerial processes. How the vehicle of tourism produces nature and society through hybrid forms of tourism mobilities, therefore, combines the fluid reality of climbing the peaks, juxtaposed with online social representations and emblematic posturing. By exploring new forms of serious leisure participation and interaction, theorists and tourism managers have an opportunity to consider how they may impact upon more 'traditional' elements of tourism.

Conclusions

This chapter has used semi-structured phenomenological interviews of Munro-baggers, informed by prior observation of online conversations and the extant literature on serious leisure, to explore classifications of participants between Munro-baggers. Emerging themes suggest four broad categories – 'hardcore', 'committed', 'moderately committed' and 'inwanderers' – although these overlap. Categories are hierarchical as prestige, status and recognition increase commensurate to one's progression from 'inwanderer' towards 'hardcore'. However, whilst some walkers aspire to 'career progression' between categories, for others the structure is non-aspirational. Most participants acknowledge loose categories, broadly agreeing on their characteristics. Most find the phenomenon benign, but some find it elitist, hegemonic or annoying. Therefore, it may reduce, increase or not affect enjoyment of the pastime, depending on the participant. Categories appear tacit, unformalized and largely affective rather than cognitive.

The characteristics informing these categories are of three strands: quantitative and relating to the central purpose of the hobby; qualitative and appertaining to Munro-bagging activities; and qualitative but *ad hominem* (i.e. appertaining to the Munro-bagger, not their activities). The characteristics with which participants judge and hierarchise themselves and each other are presented more plainly in enthusiasts' websites than was the case when Munro-bagging was situated within its original temporospatial boundaries, limited to time spent 'on the hill'. Moreover, the internet has turned a relatively solitary and geographically remote hobby into one with extensive opportunities for participation within, or at least membership of, social groups, in which members may display their credentials, scrutinise those of others, and interact. Therefore, the removal of time- and location-specificity, and the socialisation of enthusiasts, has facilitated and encouraged such hierarchising and changed the nature of participants' perceptions of self and others in the contexts of their pastime. The research offers deep insights into a pastime of increasing socio-economic importance within Scotland, and the growing influence of online spaces in shaping participant perceptions of self and others.

Finally, we recognise study limitations such as the lead researcher's inability to 'escape' membership of the study population, and its attendant subjectivism. Munro-bagging is niche, and not necessarily representative of the broader nature walking industry. For example, peak-bagging is not UK-specific, but is nuanced by location. Future investigation of these cultural and contextual differences could help to extend the debate.

Highlights

- ☐ Munro-baggers informally hierarchise themselves and each other according to the perceived 'seriousness' of their leisure.
- ☐ 'Seriousness' is conveyed through a mixture of conscious and unconscious actions and interactions.
- ☐ Pastime-specific social media sites have increased the ease, speed and frequency with which Munro-baggers convey seriousness, and the audience size.
- ☐ Social media has extended the temporospatial boundaries of the pastime beyond the hill and the walk, raising questions regarding how this elongated engagement influences serious leisure tourism.

Questions

Fort William, a small town but the main urban centre in Scotland's West Highlands, styles itself the 'Outdoor Capital of the UK'. It is adjacent Ben Nevis, Britain's highest mountain (and Munro), which attracts many ascents from serious mountaineers and casual day-trippers alike.

1. If Fort William's tourism marketers segmented potential holiday-makers into categories including 'dedicated Munro-bagger' and 'occasional hillwalker', what characteristics might they attach to each of those two segments? (Think about the behaviours and practices of each.)
2. If using an online strategy specifically to target 'dedicated Munro-baggers', how might the tourism marketers find and attract them? (Think about choices of website, messages, language and imagery.)
3. How might tourism marketers encourage 'dedicated Munro-baggers' to spread positive messages about Fort William to each other through Word of Mouth, both offline and online?

References

Brown, C.A. (2007). The Carolina shaggers: Dance as serious leisure. *Journal of Leisure Research, 39*(4), 623-647.

Chayko, M. (2008). *Portable Communities: the social dynamics of online and mobile connectedness.* Albany, NY: SUNY Press.

Cova, B., Kozinets, R.V., & Shankar, A. (Eds.). (2007). *Consumer Tribes*. London: Routledge.

Dilley, R.E., & Scraton, S.J. (2010) Women, climbing and serious leisure. *Leisure Studies* 29(2), 125-141.

Dumazedier, J. (1974). *Sociology of Leisure*. London: Elsevier Scientific

Elkington, S., & Stebbins, R.A. (2014). *The serious leisure perspective: An introduction*. Abingdon: Routledge.

Gallant, K., Arai, S., & Smale, B. (2013). Serious leisure as an avenue for nurturing community. *Leisure Sciences,* 35(4), 320-336.

Green, E. (1998). 'Women doing friendship': An analysis of women's leisure as a site of identity construction, empowerment and resistance. *Leisure Studies,* 17(3), 171-185.

Guest, G., Bunce, A., & Johnson, L (2006). How many interviews are enough? An experiment with data saturation and variability. *Field Methods,* 18(1), 59-82.

Heddon, D. & Turner, C. (2012). Walking women: shifting the tales and scales of mobility. *Contemporary Theatre Review,* 22(2), 224-236.

Heidari, K., Heydarinejad, S., Saffari, M. & Khatibi, A. (2019). Investigating the leisure behavior of Iranians: the structural model of serious leisure, recreation specialization and place attachment. *Journal of Policy Research in Tourism, Leisure and Events,* 13, 77-93.

Hollett, T., Phillips, N.C., & Leander, K.M. (2017). Digital geographies. *Handbook of Writing, Literacies, and Education in Digital Cultures,* 148-160.

Iwasaki, Y., & Havitz, M.E. (2004). Examining relationships between leisure involvement, psychological commitment and loyalty to a recreation agency. *Journal of Leisure Research,* 36, 45–72.

King, N., & Horrocks, C. (2010). An introduction to interview data analysis. *Interviews in Qualitative Research,* 142-174.

Lee, K., & Ewert, A. (2019). Understanding the motivations of serious leisure participation: A self-determination approach. *Annals of Leisure Research,* 22(1), 76-96.

Lyu, S.O., & Oh, C. (2015). Bridging the conceptual frameworks of constraints negotiation and serious leisure to understand leisure benefit realization. *Leisure Sciences,* 37, 176–193.

Møller, K., & Robards, B. (2019). Walking through, going along and scrolling back: Ephemeral mobilities in digital ethnography. *Nordicom Review,* 40(1), 95-109.

Rojek, C. (2000). *Leisure and Culture*. Basingstoke, England: Macmillan

Scottish Mountaineering Club (2018). *The Scottish Mountaineering Club Journal 2018*. Glasgow: SMC Publishing.

Shaw, S.M. (2001). Conceptualizing resistance: Women's leisure as political practice. *Journal of Leisure Research*, 33(2), 186-201.

Shen, X.S., & Yarnal, C. (2010). Blowing open the serious leisure-casual leisure dichotomy: What's in there? *Leisure Sciences*, 32, 162–179

Stebbins, R.A. (1982) Serious leisure: a conceptual statement. *Sociological Perspectives*, 25(2), 251–272.

Stebbins, R.A. (2007). *Serious Leisure: A perspective for our time*. New Brunswick, NJ: Transaction.

Stebbins, R.A. (2012) Comment on Scott: recreation specialization and the CL-SL continuum. *Leisure Services: An Interdisciplinary Journal*, 34(4), 372-374.

Stebbins, R.A. (2017). Introduction: Leisure's legacy — challenging the common-sense view of free time. In R.A. Stebbins (Ed.), *Leisure's Legacy* (pp. 1–6). Cham: Palgrave Macmillan.

Taylor, J., & Kay, S. (2015). The construction of identities in narratives about serious leisure occupations. *Journal of Occupational Science*, 22(3), 260-276.

Unruh, D.R. (1980). The nature of social worlds. *Pacific Sociological Review* (now *Sociological Perspectives*), 23, 271–296

Veal, A.J. (2017). The serious leisure perspective and the experience of leisure. *Leisure Sciences: An Interdisciplinary Journal*, 39 (3), 205-223.

Wearing, B. (1998). *Leisure and Feminist Theory*. London: Sage.

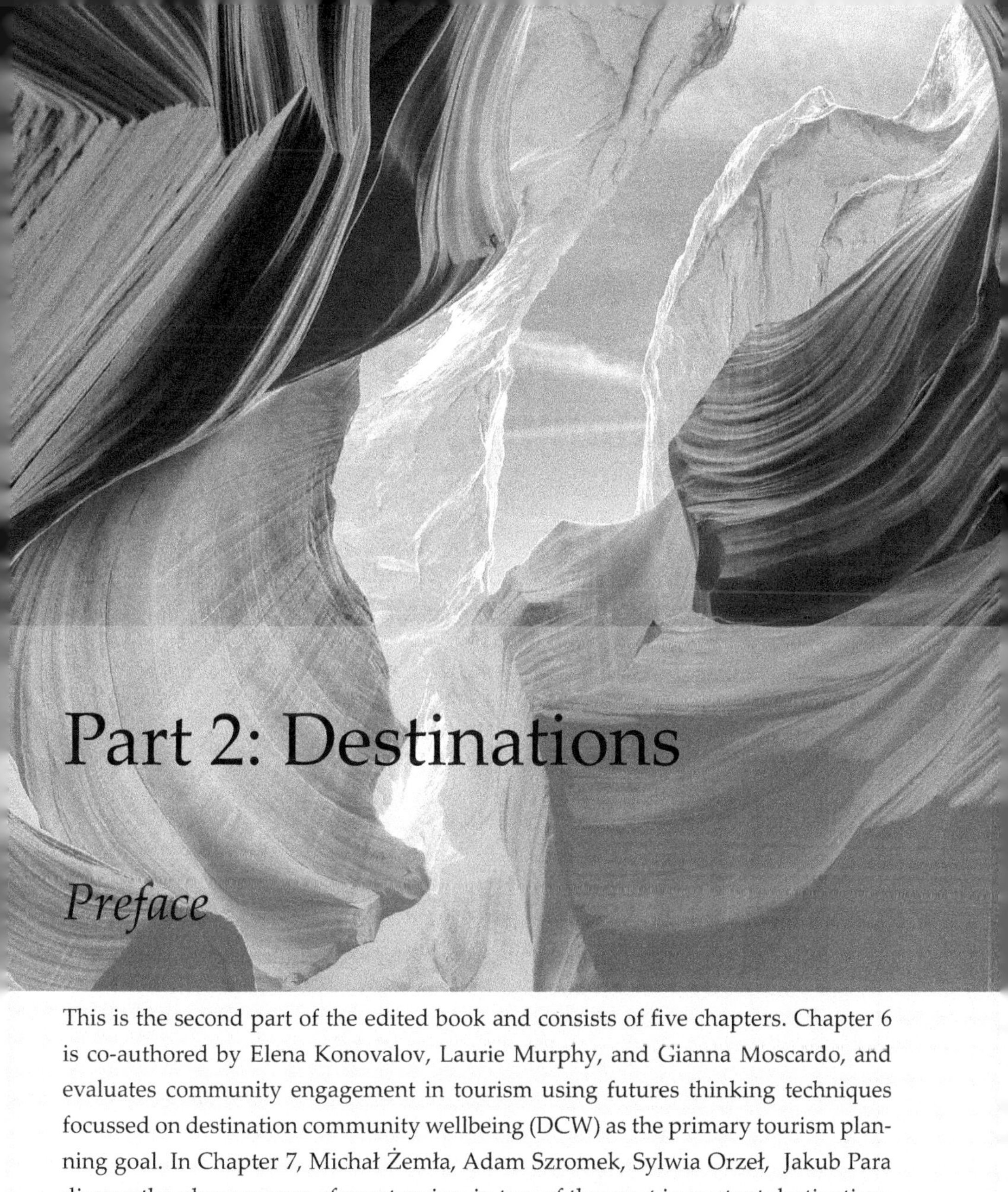

Part 2: Destinations

Preface

This is the second part of the edited book and consists of five chapters. Chapter 6 is co-authored by Elena Konovalov, Laurie Murphy, and Gianna Moscardo, and evaluates community engagement in tourism using futures thinking techniques focussed on destination community wellbeing (DCW) as the primary tourism planning goal. In Chapter 7, Michał Żemła, Adam Szromek, Sylwia Orzeł, Jakub Para discuss the phenomenon of overtourism in two of the most important destinations in Poland, (Cracow and Białka Tatrzańska). Stella Kladou, Maria Psimouli, Georgios Skourtis, Antonios Giannopoulos, and Nilgün Handan Eskitark determine the importance of specific dark tourism motives (namely those relevant to war and terrorism) for dark tourism development in Chapter 8. Chapter 9 is authored by Erisher Woyo and analyses government initiatives' role in tourism competitiveness by using data collected from suppliers operating in a distressed tourist destination. Finally, in Chapter 10 Christina Karadimitriou and Eleni Michopoulou discuss opportunities of using Extended Reality (XR) in tourism, and provide contemporary examples of XR applications.

6 Transforming tourism governance: Futures thinking for destination community well-being

Elena Konovalov, Gianna Moscardo and Laurie Murphy

Abstract

There have been growing calls for greater community engagement in tourism planning and governance. These are often accompanied by arguments for the use of more futures thinking to deal with the complex problems that tourism faces. This suggests that there is a need for different approaches to engaging destination communities in tourism planning. The aim of the present research was to evaluate such a different approach using futures thinking techniques focussed on destination community wellbeing (DCW) as the primary tourism planning goal. The outcomes of a series of community workshops that used the new futures thinking DCW approach were compared to the strategies described in relevant traditional tourism planning documents. This evaluation revealed that the residents in the futures and community wellbeing workshops generated more varied, specific, innovative and sustainable tourism futures than the traditional tourism planning processes. Tourism governance needs to focus more on empowering destination residents and on making specific links between aspects of tourism and changes in aspects of DCW.

Keywords:

Destination community wellbeing, destination resident QoL, tourism planning, tourism governance, futures thinking, community engagement

Introduction

A common inspirational quote in many contemporary social media feeds is that small changes can make big differences. Tourism has always been a dynamic phenomenon and in recent years changing global tourism systems and rising concerns about sustainability have radically altered power structures in tourism. This chapter argues that two critical changes in tourism governance have the potential to transform tourism allowing for more effective responses to the dynamic conditions it operates under. The first change is explicitly shifting the primary goal of tourism planning from increasing tourist numbers and expenditure to using tourism as a tool to develop destination community well-being (DCW). The second change is the implementation of futures thinking in destination community engagement in tourism governance. The chapter reports on an evaluation of a DCW approach using futures thinking to examine tourism governance in three regional destinations in Australia. The proposed new process involved a focus on community well-being and the use of futures thinking in a series of activities with community stakeholders designed to generate new tourism futures.

The chapter begins by briefly reviewing recent innovations in public participation in planning, integrating futures thinking into planning and using a DCW approach to tourism planning. It then describes the futures thinking DCW strategy used in three Australian destinations to generate ideas for tourism development and governance. After an overview of the evaluation methodology, the outcomes of the strategy were compared to ideas generated by the relevant destination marketing/management organizations (DMO) through traditional tourism planning processes. The resulting differences demonstrate the transformations possible through this new approach and how these can improve tourism flexibility to respond to various dynamic forces.

Challenges to transforming tourism governance for destination communities

In the years prior to the 2020 COVID-19 pandemic there were increasing discussions about the need to radically rethink the nature of tourism. There have also been calls for tourism to focus more seriously on its relationship to both sustainability in general, and the wellbeing of destinations in particular (Dwyer, 2018). These calls reflect concerns from destination residents about tourism's negative impacts, issues with tourism's contributions to global carbon emissions and criticism of outdated, overly simplistic approaches to tourism governance. Two common themes in these calls for new paradigms are:

- the need for a stronger focus on destination resident engagement and empowerment in tourism planning and governance, and
- a shift in the goals of tourism planning from increasing tourist numbers and tourism business success, to generating benefits and value to destination communities (Dangi & Jamal, 2016; Dwyer, 2018; Moscardo & Murphy, 2014; Saarinen et al., 2017).

An examination of the wider literature on planning in general, and community development planning more specifically, identified five key trends:

- The rise of citizen power and a shift towards planning being primarily about citizen empowerment;
- The increasing use of new technologies to allow for the expression of that citizen power;
- A shift from technical, rationalist planning paradigms to more ethical and value-based frameworks that reflect this rise in citizen power;
- Greater flexibility, creativity and diversity in techniques for engaging with stakeholders; and
- Recognition of the need to understand cultural differences in attitudes towards governance (Gaynor, 2013; Glackin & Dioniso, 2016; Green & Haines, 2012; Groves et al., 2013; Howlett et al., 2015; Johnson et al., 2016; Moscardo, 2019; Munster et al., 2017).

Moscardo (2019) argues that these trends are not evident in the tourism planning and development literature. Studies of destination resident engagement in tourism planning continue to report that it is rarely included in tourism planning practice (Bello et al., 2016; Idziak et al., 2015; Mak et al., 2017; Pforr & Brueckner, 2016). On those few occasions when destination residents are included in the process it is mostly at the lower levels of Arnstein's (1969) ladder of community participation. These efforts exist in the token consultation and non-participative information provision categories of Marzuki and Hay (2013). The two main directions emerging from these five trends are a specific focus on planning to improve the wellbeing of the target communities and more extensive use of futures thinking. It is suggested that these community engagement processes focus on envisioning and generating alternative futures scenarios and solutions (Glackin & Dioniso, 2016; Green & Haines, 2012; Johnson et al, 2016).

Tourism is often described as an instrument for community development, particularly in regional communities (Saarinen, 2019). If the goal of community development is empowering and improving conditions for community residents (Kenny, 2016), then arguably the primary goal of tourism planning is to find ways to improve the wellbeing of destination community residents. While this is consistent with the trends above, it is rare to find tourism planning practices guided by the goal of directly enhancing different aspects of DCW (Dangi & Jamal, 2016; Dwyer, 2018; Saarinen et al., 2017). New tourism planning frameworks guided by DCW are emerging and they provide examples of how tourism governance can be refocussed to see tourism as tool for DCW. See frameworks suggested by Moscardo and Murphy (2014) and Shoeb-Ur-Rahman and colleagues (2020) for examples of a DCW approach to tourism planning.

Futures thinking can be described as a mindset focussed on understanding patterns and processes of change so that current planning can be altered to avoid potential problems and enhance desirable outcomes (Varum et al, 2011). Futures thinking is about using different techniques and processes to encourage people to consider options beyond *"their traditional operational and conceptual comfort zone"* (Amer et al., 2012, p. 25). The most common technique to support futures thinking is the development and analysis of alternative future scenarios (Burnam-Fink, 2015). Like community engagement in planning, there has been very little use of futures thinking in tourism (Yeoman & Postma, 2014).

Even before the COVID-19 global pandemic, tourism was facing challenges related to changing power structures in destinations, the impacts of new technologies and pressures to respond to sustainability challenges. The theme of this edited volume recognises these pressures and the need to find ways to transform underlying systems and practices to make tourism more flexible and dynamic. This chapter is focussed on these issues as they apply to tourism governance at the destination level.

The literature review identified the need for greater community engagement in tourism planning and governance and for the use of more futures thinking to deal with the complex problems that tourism faces regarding sustainability. There is a clear need for a change in direction in how tourism practitioners engage destination communities in tourism planning. This chapter therefore has two aims. The first is to describe a tourism planning and development based on DCW as the primary tourism planning goal and using futures thinking techniques. The second is to report on a qualitative evaluation of this Futures Thinking DCW approach.

A futures thinking DCW approach to engaging destination residents in tourism governance

The proposed futures thinking DCW planning approach focusses on understanding links between tourism and the various capitals that make up community wellbeing (Emery & Flora, 2006; Pigg et al., 2013). The process was guided by the first five of Inayatullah's (2008) six pillars approach to futures thinking. Table 6.1 describes these five pillars and links them to the steps taken in this new community engagement process.

Table 6.1: Matching Inayatullah's six futures thinking approach to the community engagement process

Pillar	Action in engagement process
Mapping: Understanding the system and its past and present states	Analysis of indicators of tourism development, measuring objective wellbeing indicators and surveying destination residents on perceived wellbeing and att itudes to future tourism
Anticipation: Identifying key issues	Focus on human and social capital, community identity and pride, and community services as key issues often excluded in tourism planning
Timing: Using tools such as futures wheels to map out changes over time **Deepening:** Exploring in more detail causal links between governance decisions and consequences	Based on the previous analyses linking aspects of tourism to impacts on community wellbeing, a workshop used futures wheels with backcasting to link longer term preferable futures to governance and development strategies in the short and medium term
Creating alternative futures	Identifying desirable tourism futures from the futures wheel workshops

The *mapping* step took a quantitative approach and used resident surveys and analysis of government indicators. These indicators measured the level and style of tourism development and the current state of overall resident wellbeing. The surveys assessed resident perceptions of the impacts of different types of tourists and attitudes towards future tourism development. More details on this element and its results can be found in Konovalov et al. (2019). In step two, *anticipation*, the key issues to focus on were identified from reviews of research into resident perceptions of tourism impacts on their communities. Reviews of tourism impacts typically suggest that considerable effort has been expended on environmental and economic impacts, but little attention has been paid to the social impacts of tourism (Konovalov et al., 2019). Reviews of research on the social impacts of tourism and resident perceptions of tourism (Deery et al., 2012; Gursoy et al., 2019) identified the following areas of community well-being that require further examination:

- ☐ Human capital: the skills and abilities of residents which include education, skills, health, and self-esteem;
- ☐ Social capital: connections among people in a community including networks and relationships;
- ☐ Community identity and pride; and
- ☐ Community services that are available to residents including public services, recreational services and public transport.

The results of the mapping analysis were widely distributed to community groups, government agencies and tourism organisations. A summary of these quantitative results formed a core element of the community workshops that covered pillars three to five in Table 6.1. These community workshops explored in more detail the links between different tourism planning and management decisions and preferable futures. The 28 participants invited to the three regional workshops were identified from lists of community stakeholders used during previous research activities in the regions. These included representatives of local councils, tourism and community organisations, tourism operators, small business owners and other interested community representatives. Workshop invitations were sent to the identified community stakeholders who were free to share them with others they thought could contribute. The role of the workshop facilitators was presented as knowledge-brokers rather than consultants to the existing power structure (Jami & Walsh, 2016).

The workshops consisted of a short presentation of research findings on tourism and the social aspects of community well-being in the region, followed by a brainstorming activity. The brainstorming activity combined a futures wheel technique with backcasting to identify ways in which tourism could positively contribute to each of the four key aspects of community wellbeing. The technique is a structured method aimed at facilitating group discussion and systematic thinking about either the future consequences or how to get to a desirable future state (Benckendorff, 2008; Bengston et al., 2020). It has previously been used to explore the links between tourism and quality of life/community well-being (Benckendorff et al., 2009; Moscardo et al., 2017). The exercise consists of placing a statement about a desirable future at the centre of the to-be-constructed wheel. Then participants are invited to contribute their thoughts and ideas on conditions that must be present for this desired future to be realised.

For the workshops the following statements were placed at the centre of the wheels: '*In 10-15 years tourism in the region will make a positive contribution to human capital [social capital/community identity and pride/ community services]*'. Thus, at the starting point of the workshops the participants had four potential wheels, one for each of the four key issues linked to community well-being. Facilitators then asked the participants what would have to happen in their community for them to be able to make such statements in 10-15 years' time, thus the meanings of those statements for each community were clarified and the first ring of the futures wheel created. Next participants were asked to think about what needs to start happening in the present for these desired future changes to occur in their community, thus creating another ring in the futures wheels. At the end of each workshop wheels for each of the social aspect of community well-being were created. These captured respondents' thoughts and ideas about desirable future tourism scenarios and suggested tourism strategies for achieving these. Reports on these workshops were compiled and presented to the relevant DMOs and other tourism organizations in each region.

Methodology

A case study methodology using archival analysis was conducted to achieve the aim of evaluating the Futures Thinking DCW approach to engaging destination residents in tourism planning. This evaluation compares the outcomes of the futures workshops conducted in three case study regions to the outcomes of the traditional tourism planning approaches in the same regions. The evaluation compared the major tourism developments and strategies outlined in the traditional planning documents to the major outcomes of the futures thinking DCW workshops. The analysis focussed on innovativeness and the specificity of links made between tourism decisions and aspects of DCW.

Sampling and sample case study characteristics

As the aim of the evaluation research was to demonstrate the value of the Futures Thinking DCW approach, it was important to find destination regions seeking to respond to changing tourism dynamics. The aim also required a diversity of community experience with tourism and its positive and negative impacts. These two criteria support the generalizability of the conclusions beyond a specific context as outlined in Tsang's (2013) discussion of improving the theoretical value of case study research. Three regional communities in Northern Australia were selected, that had varying styles and scales of tourism development and were engaged in planning processes. These included a well-established coastal tourism destination (Region A), an area with a long established but low intensity tourism sector (Region B), and a region with an emerging but still small tourism sector (Region C). Table 6.2 provides a summary of the three case study destinations based on available statistics on numbers and types of tourists, available accommodation and history of tourism development.

Archival analysis

Archival analyses, or the qualitative coding of existing documents or text-based materials, are often used in case study methodologies (Ventresca & Mohr, 2017). The data for the archival analysis in this evaluation came from two sources. The first was the reports produced by the workshop facilitators. The second was the relevant destination plans and tourism investment opportunity strategies published for the three regions at the same time as the workshops were conducted. These documents were subjected to content analysis and thematic coding following the steps described by Denzin and Lincoln (1994).

Table 6.2: Summary of the three case study destinations

Features	Region A Well-established intense tourism	Region B Long established Low intensity tourism	Region C Emerging tourism
Major economic activities	Tourism	Agriculture	Agriculture & mining
Geography	Coastal	Inland mountainous region adjacent to a major coastal resort area	Coastal adjacent to a major coastal resort area
Approximate population: Annual tourist visit	11,000: 2.8 million	45,000:1.4 million	9,000: 0.5 million
Tourist types	30% International, 35% domestic from outside the region	4% International 21% domestic from outside the region	6% international, 16% domestic from outside the region
Typical length of stay	1 week	65% day visitors from adjacent region; overnight visitors spend either 1- 4 days or 31 days or more	1-4 days with nearly half day visitors from the adjacent region
Accommodation	Large scale resorts and serviced holiday apartments	Caravan/RV parks Medium motels Holiday cottages	Caravan/RV parks and small motels

Results and discussion

The workshops at each community generated lively discussions about future tourism development options and how tourism could be used to enhance the community wellbeing. The participants' main thoughts and ideas were captured in the generated futures wheels. An example can be seen in Figure 6.1. The figure identifies issues related to community pride in the region, including concerns about empty shop fronts and the appearance of public space. It also includes opportunities to enhance community pride through festivals and events and enhancing culture and arts. Suggestions for using tourism to support these enhancers are then listed as well as ways in which tourism might be able to help solve the problems. At the outermost points of the wheel are suggestions for immediate tourism strategies that could be implemented. These included the development of camping facilities, attracting conservation volunteer tourists, greater tourism sponsorship of local events, and more innovation in the way governments advertise and award development funds.

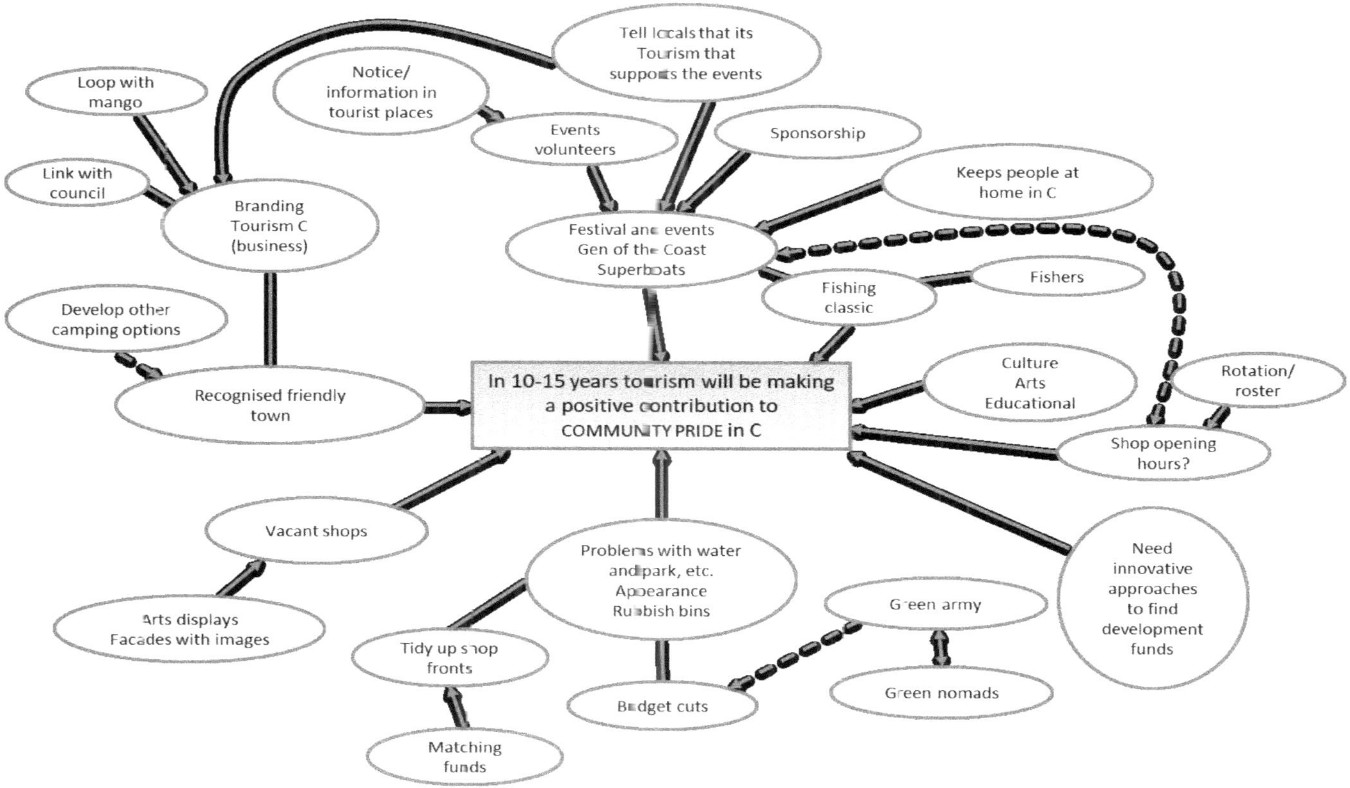

Figure 6.1: A futures wheel for Region C focussed on how future tourism could contribute to community pride.

Table 6.3 summarises the main suggested tourism future development strategies that emerged from all the workshop futures wheels organised by the four community wellbeing elements. Common themes emerged across the three regions in all four elements of community wellbeing. The specifics of these themes did however differ significantly across the different contexts. While all three regions saw tourism as a tool for employment, Region A sought more professionals and Region B wanted more full-time positions for young people. Region C wanted employment targeted specifically to locals. Similarly, all three wanted tourism to encourage new residents to migrate to the region, adding to the population base. Region A, however, specifically sought families, Region B sought young people and Region C desired older amenity migrants/retirees. In terms of tourism and community services, participants from Region A wanted future tourism to focus on more recreation and entertainment opportunities for families. Participants from Region B thought that developing medical and wellbeing tourism could support improvement of health services for the region. Participants in Region C were much more focussed on increasing local retail options and better transport services to support amenity migrants.

Table 6.3: Summary of suggested tourism future development strategies

Focus	Common theme
Human capital	Tourism offers employment Tourism could support education Tourism could be used to attract new residents
Social capital	Encourage tourists to engage in social activities with residents Build connections between tourism and other sectors, especially with agriculture Encourage domestic tourism from adjacent regions to increase opportunities for local businesses to connect regionally Find ways to support tourists acting as volunteers adding skills to the local community
Community identity & pride	Develop tourism brand linked to local resident perceptions of community identity and culture Involve more local community groups in tourism activities Encourage attendance by locals and tourists to events
Community services	Use increased demand from tourists to increase and improve health, retail and recreation services and opportunities

The Futures Wheel backcasting activities encouraged participants to think about how tourism could be used to enhance different aspects of community wellbeing and what tourism actions and strategies were required. The process also generated a set of future tourism scenarios that included forms of tourism, desirable tourist markets and specific tourist activities/experiences that could be developed. These are summarised in Table 6.4 and contrasted with the future tourism scenarios described in the documents developed using traditional tourism planning approaches.

Table 6.4: Future tourism scenarios

Region	Workshop tourism futures	Traditional planning tourism futures
A (established large)	More farm tourism More land based eco-tourism Increase tourist activities based on history of the region More older retired domestic tourists	More nature based (especially marine) tourism More activities for families based on nature More events More international tourists More domestic families
B (established small)	Health and wellbeing destination with medical tourism Support for local businesses to use new technologies More home stay experience Educational tourism presented by locals More domestic tourists (especially young people)	Health & wellness tourism a possible future direction More eco-lodges & outdoor adventure More commercial all-inclusive resort style accommodation Educational tourism in established institutions More older domestic tourists and more tourists from China and India
C (emerging adjacent to a large destination)	Local food-based tourist experiences and events including cooking classes and celebrity chefs Tourist experiences based on local history and artists More innovation in funding locals to offer tourism including crowdsourcing More domestic tourists	More nature-based experiences A local produce centre to showcase regional agriculture More external grants More domestic short break tourists More international tourists especially from China

While there was some overlap between the two sets of tourism futures summarized in Table 6.4 there were also substantial differences both in content and approach. The tourism planning documents analysed were based on governance processes very similar to those outlined by Moscardo (2011). The opportunities for destination community residents to participate in tourism development and planning decisions were limited to commenting on plans and futures already developed by the relevant organisations. For example, all the formal tourism planning documents mentioned stakeholder engagement but the lists of stakeholders, where provided, were exclusively tourism businesses, tourism organizations and some government departments. None of the documents reported engaging with the wider community. In all the documents the stakeholder engagement that was conducted played only a minor role in the planning process. Most of the documents relied primarily on analyses of recent tourism, forecasts of large scale, usually global, tourism trends and suggestions proposed by external large tourism developers/investors.

In all the planning documents the goals of the plans and strategies were to specifically increase tourist numbers, length of stay and expenditure. Saarinen, Rogerson and Hall (2017) describe such goals as typical of boosterism approaches to tourism planning and argue they are unlikely to generate sustainable tourism. Boosterism approaches are also characterised by conceptualising destination communities as resources to be used to increase tourism and do not consider tourism as a tool for destination community development. This was evident in the present set of documents with one specifically stating that the key to their success in achieving their tourism growth goals would *"be engaging the community as a resource"*. Another calling for governments to *"ensure community and service infrastructure requirements keep pace with a growing population so as to support sustainable tourism growth"*. As tourism is not seen in these documents as a community development tool, specific links between tourism and its impacts on aspects of DCW were never discussed. All the documents assumed tourism was desirable because it created employment and generated revenue, although no specific details on how this would be achieved were ever provided.

This lack of specificity was also evident in many of the other dimensions of the tourism futures described in the plans. So, for example, one planning document noted that health and wellness tourism was a possible tourism future for Region B, which was consistent with what the community workshop suggested. The description of this tourism future was, however, very vague with no specific actions listed to move towards this goal. Similarly, the planning documents for Region C mentioned the need for increased external grants to fund tourism development projects which was consistent with one outcome of the community workshops. Again, the need was identified as a key strategy in the planning documents, but no specific actions were identified to achieve this. This lack of specific links and actions is of concern as the planning documents were based around short time frames which should make it easier to be more specific. Such a lack of *"robust and detailed"* strategies for achieving tourism planning goals has also been noted by other researchers (McLoughlin & Hanrahan, 2016, p. 33). Overall, the Futures Thinking DCW approach generated a wider range of issues, solutions and more specific and concrete links between different aspects of tourism and DCW.

The traditional plans were also much less innovative than the scenarios and options generated in the community workshops. The traditional tourism plans tended to look backwards at past patterns and many of the strategies were extensions of what had been done in the past. Alternatively, they adopted ideas used in well-known competitor destinations. Not surprisingly the traditional plans were very similar in their contents and thus described destination images and experiences that were very similar to each other. Dwyer (2018) argues that this traditional approach needs to change as the creation of unique places is what is required to avoid negative competition between destinations.

Conclusions

The Futures Thinking DCW focussed workshops generated more detailed, concrete and innovative tourism planning strategies and actions than did the traditional tourism plans for the three destinations studied. These outcomes were very similar across all three destinations even though they differed considerably in style, extent, history and dependence upon tourism. It is important to note that these conclusions were based on only one workshop conducted in each region. Arguably this may have limited the range of alternative tourism futures and does not allow for assessment of wider support for the proposed tourism futures. It should be noted, however, that even with only one workshop in each place, the outcomes were more innovative, extensive and specific than those created through traditional planning processes. While it is not possible from the present evidence to argue that the specific tourism futures described above should be implemented, the process generates better outcomes than traditional approaches. The obvious next steps in this community wellbeing future thinking approach would be to:

- ☐ Conduct more workshops to generate a wider range of scenarios and strategies;
- ☐ Present these resulting tourism futures to the whole community for comment;

- ☐ Assess them in terms of sustainability and market and business viability;
- ☐ Implement them; and
- ☐ Evaluate the outcomes in terms of improvements to community wellbeing.

Inayatullah's (2008) final pillar for futures thinking is *transformation* or the implementation of the actions required to begin moving towards the desirable futures. The present study was unable to take these specific workshop outcomes into any formal planning or implementation processes. The group from Region B were, however, able to progress several of the specific ideas through their personal networks. The group from Region C, by way of contrast, were much more pessimistic about their capacity to implement their ideas. This diversity of response presents both a challenge to the idea of transformation in tourism governance and an area for further research. Studies are needed into the factors that influence the extent to which destination residents feel able to change tourism. Research is also needed into the willingness of DMOs and relevant government agencies to accept and embrace a greater role for residents in tourism governance.

The present volume seeks to understand the dynamics of tourism systems and how tourism practices can be adapted so that tourism can respond to these pressures and changes. The results of this evaluation demonstrate the potential to change tourism governance practice to support more resident engagement in destination governance and greater innovation in tourism planning. The use of futures thinking offers a new approach to tourism planning in general and the effective engagement of destination residents in tourism governance. There has been considerable discussion in various academic fora on the potential of the COVID 19 pandemic to transform tourism towards greater sustainability. Unfortunately, there has been little concrete evidence of how this change might happen. In the present case the shift of DMOs' focus from international to domestic tourism has encouraged a greater interest in resident views on tourism futures. It has also generated a greater willingness to try alternative governance approaches. Thus, there is an important opportunity to expand the use of techniques such as those described in the present chapter to transform tourism governance.

Highlights

- ☐ This chapter describes a new approach to engaging destination communities in tourism planning based on futures thinking and enhancing community wellbeing.
- ☐ It compares the outcomes of this new approach to outcomes from traditional planning approaches, demonstrating that residents can generate more varied, specific and sustainable tourism futures.
- ☐ Tourism governance needs to focus more on empowering destination residents and on making specific links between aspects of tourism and changes in aspects of DCW.

Questions

1. What is needed in tourism education to support greater use of community engagement tools and futures thinking in tourism governance?
2. Do you know of, or can you find, an example of good practice in destination community engagement in tourism governance and explain why you think it is good practice?
3. Can you suggest other ways to encourage greater creativity in the way tourism planners and destination managers approach the challenge of using tourism as a tool for improving destination resident quality of life or wellbeing?

References

Amer, M., Daim, T. & Jetter, A. (2013) A review of scenario planning, *Futures*, 46, 23-40.

Arnstein, S.R. (1969) A ladder of citizen participation, *Journal of the American Institute of Planners*, 35(4), 216-224.

Bello, F.G., Carr, N. & Lovelock, B. (2016) Community participation framework for protected area-based tourism planning, *Tourism Planning & Development*, 13(4), 469-485.

Benckendorff, P. (2008) Envisioning sustainable tourism futures, *Tourism and Hospitality Research*, 8(1), 25-36.

Benckendorff, P., Edwards, D., Jurowski, C., Liburd, J. J., Miller, G. & Moscardo, G. (2009) Exploring the future of tourism and quality of life, *Tourism and Hospitality Research*, 9(2), 171-183.

Bengston, D. N., Westphal, L. M. & Dockry, M. J. (2020) Back from the future, *World Futures Review*, 12(3), 270-278.

Burnam-Fink, M. (2015) Creating narrative scenarios, *Futures*, 70, 48-55.

Dangi, T. B. & Jamal, T. (2016) An integrated approach to 'sustainable community-based tourism', *Sustainability*, 8(5), doi:10.3390/su8050475.

Deery, M., Jago, L. & Fredline, L. (2012) Rethinking social impacts of tourism research, *Tourism Management*, 33(1), 64-73.

Denzin, N. K. & Lincoln, Y. S. (1994) Introduction: Entering the field of qualitative research, in N. K. Denzin & Y. S. Lincoln (eds.), *Handbook of Qualitative Research*, Thousand Oaks: Sage, pp. 1-17.

Dwyer, L. (2018) Saluting while the ship sinks: the necessity for tourism paradigm change, *Journal of Sustainable Tourism*, 26(1), 29-48.

Emery, M. & Flora, C. (2006) Spiraling-up: Mapping community transformation with community capitals framework, *Community Development*, 37(1), 19-35.

Gaynor, N. (2013) The tyranny of participation revisited, *Community Development Journal*, 49(2), 295-310.

Glackin, S. & Dionisio, M.R. (2016) ''Deep engagement' and urban regeneration', *Land Use Policy*, 52, 363-373.

Green, G.P. & Haines, A. (2012) *Asset Building and Community Development*, 3rd edn, Los Angeles: Sage.

Groves, C., Munday, M. & Yakovleva, N. (2013) Fighting the pipe: Neoliberal governance and barriers to effective community participation in energy infrastructure planning, *Environment & Planning C*, 31, 340-356.

Gursoy, D., Ouyang, Z., Nunkoo, R. & Wei, W. (2019) Residents' impact perceptions of and attitudes towards tourism development, *Journal of Hospitality Marketing & Management*, 28(3), 306-333.

Howlett, M., Mukherjee, I. & Woo, J.J. (2015) From tools to toolkits in policy design studies, *Policy & Politics*, 43(2), 291-311.

Idziak, W., Majewski, J. & Zmyślony, P. (2015) Community participation in sustainable rural tourism experience creation, *Journal of Sustainable Tourism*, 23(8-9), 1341-1362.

Inayatullah, S. (2008) Six pillars: futures thinking for transforming, *Foresight*, 10(1), 4-21.

Jami, A. A., & Walsh, P. R. (2016) Wind power deployment: The role of public participation in the decision-making process in Ontario, Canada, *Sustainability*, 8(8), doi:10.3390/su8080713.

Johnson, M.L., Bell, K.P. & Teisl, M.F. (2016) Does reading scenarios of future land use changes affect willingness to participate in land use planning?, *Land Use Policy*, 57, 44-52.

Kenny, S. (2016). Community development today: engaging challenges through cosmopolitanism?. *Community Development Journal*, 51(1), 23-41.

Konovalov, E., Murphy, L. & Moscardo, G. (2019) An exploration of links between levels of tourism development and impacts on the social facet of residents' quality of life, in A.M. Campón-Cerro, J.M. Hernández-Mogollón & J.A. Folgado-Fernández (eds.), *Best Practices in Hospitality and Tourism Marketing and Management,* Cham: Springer, pp. 77-107.

Mak, B.K.L., Cheung, L.T.O. & Hui, D.L.H. (2017) Community participation in the decision-making process for sustainable tourism development in rural areas of Hong Kong, China, *Sustainability*, 9, 1695-1708.

Marzuki, A. & Hay, I. (2013) Towards a public participation framework in tourism planning, *Tourism Planning & Development*, 10(4), 494-512.

McLoughlin, E. & Hanrahan, J. (2016) Local authority tourism planning in Ireland, *Journal of Policy Research in Tourism, Leisure and Events*, 8(1), 33-52.

Moscardo, G. (2011) Exploring social representations of tourism planning, *Journal of Sustainable Tourism*, 19(4-5), 423-436.

Moscardo, G. (2019) 'Rethinking the role and practice of destination community involvement in tourism planning', in K. Andriotis, D. Stylidis & A. Weidenfeld (eds.), *Tourism Policy and Planning Implementation*, London: Routledge, pp. 36-52.

Moscardo, G., Konovalov, E., Murphy, L., McGehee, N. G. & Schurmann, A. (2017) Linking tourism to social capital in destination communities', *Journal of Destination Marketing & Management*, 6(4), 286-295.

Moscardo, G. & Murphy, L. (2014) There is no such thing as sustainable tourism, *Sustainability*, 6 (5), 2538-2561.

Munster, S., Georgi, C., Heijne, K., Klamert, K., van der Meer, H. (2017) How to involve inhabitants in urban design planning by using digital tools?, *Procedia Computer Science*, 112, 2391-2405.

Pforr, C. & Brueckner, M. (2016) The quagmire of stakeholder engagement in tourism planning, *Tourism Analysis*, 21(1), 61-76.

Pigg, K., Gasteyer, S. P., Martin, K. E., Keating, K. & Apaliyah, G. P. (2013) The community capitals framework, *Community Development*, 44(4), 492-502.

Saarinen, J. (2019) Communities and sustainable tourism development, in S. McCool & K. Bosak (eds.), *A Research Agenda for Sustainable Tourism*, Cheltenham: Edward Elgar, pp. 206-222.

Saarinen, J., Rogerson, C. M. & Hall, C. M. (2017) Geographies of tourism development and planning, *Tourism Geographies*, 19(3), 307-317.

Shoeb-Ur-Rahman, M., Simmons, D.G., Michael Shone, M. & Ratna, N. (2020) Co-management of capitals for community wellbeing and sustainable tourism development: A conceptual framework, *Tourism Planning & Development* 17(2), 225-236.

Tsang, E. W. (2014) Generalizing from research findings: The merits of case studies, *International Journal of Management Reviews*, 16(4), 369-383.

Varum, A., Melo, C., Alvarenga, A. & Soeiro de Carvalho, P. (2011) Scenarios and possible futures for hospitality and tourism, *Foresight*, 13 (1), 19-35.

Ventresca, M.J. & Mohr, J.W. (2017) Archival research methods, in J. Baum (ed.), *The Blackwell Companion to Organizations* (pp.805-828). Oxford: Blackwell.

Yeoman, I. & Postma, A. (2014) Developing an ontological framework for tourism futures, *Tourism Recreation Research*, 39(3), 299-304.

7 The costs and benefits of overtourism – perspectives of residents of cities and resorts

Michał Żemła, Adam R. Szromek, Sylwia Orzeł and Jakub Para

Abstract

The costs and benefits of tourism development and their perception by destinations' residents have been a subject of research for many years. Although relevant studies considered the effects of overtourism, resident perceptions of overtourism impacts were not analyzed in a comparative context of different types of destinations. Large cities, for example, require different tourism development approaches and tools as they differ from other destinations, especially tourism resorts. This study therefore examines residents' perceptions of overtourism by focusing on two important yet different destinations in Poland, Cracow and Białka Tatrzańska. The results confirm significant differences in how residents perceive the impact of overtourism with the biggest difference being evident on how the benefits of tourism development are perceived. The chapter draws conclusions that can contribute significant insights on the management of tourism in different types of destinations.

Key words:

Overtourism; urban tourism; costs and benefits of tourism; social impact of tourism

Introduction

Large cities are becoming more and more popular as tourist destinations (Namberger et al., 2019). In fact, the 21st century represents the period of rapid development of what has become known as 'urban tourism' or 'city tourism' (UNWTO, 2018; Maitland, 2018). This sharp increase in urban tourism demand was due to various reasons: a) people taking shorter holidays but traveling more often; b) low-cost carriers allowing people to reach cities at affordable prices; and c) cities become more attractive by organizing various events as well as becoming increasingly popular as locations for culture, shopping and sightseeing (Namberger et al., 2019). Even so, research on urban tourism began only in the 1990s. Up to that point, tourism played a marginal role in the development of cities and the monuments and places attractive to tourists were not recognized as key tourism products (Christaller, 1964). Ashworth (1989, 1992) pointed to the significant scientific negligence of the interface of urban cities and tourism development, calling for more research on the growing urban tourism phenomenon. The last decade of the 20th century, thus, marked the beginning of an intense international and interdisciplinary academic debate on various aspects related to urban tourism development. Despite the significantly increasing amount of literature, Ashworth (2003) indicated that scientific knowledge about the development of urban tourism remains insufficient, especially with regard to resident perceptions.

Residents' support for tourism development is strongly related to its success (Gürsoy, et al., 2002; Su et al., 2018). Considering that tourism is acknowledged as one of the most sustainable economic growth strategies for cities (Koens et al., 2018), one would expect that city residents will perceive tourism favorably and extend support to tourism development (Goodwin, 2017). Even so, strong anti-tourism attitudes and movements are currently emerging among residents of large cities, mostly as a result of the impacts of overtourism. As the phenomenon of overtourism and in parallel 'tourismphobia' came into being (Milano, 2017; Gürsoy, 2019), it is becoming evident that the relationship between the success of a destination and residents' support may be negative, at least in the case of large city destinations. Overtourism is most evident in cities that have achieved large market success as a result of a great number of tourists, accompanied by significant income from tourism. The consequences of overtourism are often so significant that sometimes it leads to as a 'third-order change' (Hall, 1994), which involves a shift in the policy paradigm including an adaptation of a new goal hierarchy (Żemła, 2020). Undoubtedly, overtourism is one of the most typical and characteristic phenomena associated with contemporary tourism. Although its rapid development is disturbed by the COVID-19 pandemic, it requires creating new approaches in terms of tourism management.

As a result of overtourism in large cities, it is necessary to perform research to uncover the conditions for sustainable urban tourism development. Tools and approaches followed so far in managing tourism development in tourist resorts are largely ineffective in the case of cities (Goodwin, 2017; Żemła, 2020). Research on the impact of tourism on the lives of residents of tourist destinations have been conducted, yet without a detailed examination of how this impact may be differentiated in destinations of various types. Although some authors (Peeters et al., 2019) indicated that overtourism is present in various destinations, others limited their research on overtourism only to metropolises (Goodwin, 2017; Milano et al., 2019; Capocchi et al., 2019). To date, the differences between how the impact of overtourism is perceived by residents of different types of destination has not been the subject of scientific analysis. This chapter attempts to fill this gap by examining how the impact of tourism development of a well-known and popular tourist resort (Białka Tatrzańska) and a large city (Cracow) in Poland is perceived by residents.

The impact of urban tourism on residents' lives

One of the social consequences of the development of urban tourism is the change of residents' attitude towards tourists. This change depends on whether and, if so, how the positive and negative effects of tourism are perceived by residents. The changes in the attitude of residents are often explained using Doxey's (1976) model, which indicates that the negative attitudes of residents towards tourists are the result of exceeding the limits of social carrying capacity. According to this model, changes in residents' attitudes to tourism development can be divided into four successive stages: from euphoria, through apathy, irritation, even to as much as conflict. While mass tourism brings significant economic gains to cities, it has a negative impact on their development as a result of overtourism which may yield various negative effects on residents' lives (Su et al., 2018).

According to Novy and Colomb (2017), the main sources of conflicts and the causes of protests against urban tourism development can be grouped into four categories: economic (e.g. commercial/retail gentrification, price increase), physical (e.g. land-use conflicts, crowds), social and sociocultural (e.g. commercialization, exploitation and distortion of culture, heritage and public space, increase in petty crime) and psychological (e.g. feeling of losing control over the future of local and city communities, loss of feeling of belonging or attachment to the community). Other authors indicated specific negative impacts of overtourism on the lives of residents, including the increase in the prices of services, housing, and real estate (Goodwin, 2017; Namberger et al., 2019; Żemła, 2020) as well as congestion in places, roads, transportation means, queues in shops and service facilities (Novy & Colomb, 2017; Milano et al., 2019). Another factor that is often presented in the literature as an overtourism problem is the improper behavior of tourists (Goodwin, 2017; Phi, 2019) which may abuse local traditions. Also, loud and/or drunk tourists may lower residents' sense of security.

Although the above-mentioned problems have been presented in the literature for years (Doxey, 1976; Pizam, 1978; Butler, 1980; Canestrelli & Costa, 1991), they have become more severe in light of the surge of tourism in cities. Likewise, the development of tourism in cities has created additional problems (Herrera et al., 2007) such as those brought by the popularity of sharing economy platforms (e.g. Airbnb, Uber). Numerous researchers (Wegmann & Jiao, 2017; DiNatale et al., 2018) confirm that the extremely fast development of Airbnb-type platforms seriously impacts the real estate market in cities. Some authors (Horn & Merante, 2017) noticed that home sharing is a factor that rises rental prices by diminishing the number of units available to potential residents. In addition to affordability problems and property shortages, the development of sharing economy platforms in tourism should generate additional income for residents, thus improving their quality of life (Horn & Merante, 2017). However, these economic benefits are often unevenly distributed, with much of them leaking outward to outside homeowners.

In addition to the typically economic consequences related to renting and housing prices, residents also face other types of problems as Airbnb rentals allow tourism to spread to districts that were previously only inhabited by residents. Offering accommodation in a district that has not yet been visited by tourists has become an attractive earning opportunity for property owners, while for tourists it has become an opportunity to learn about how local people live. Many visitors, attracted by the Airbnb 'live like a local' catchphrase, use this alternative accommodation option (Goodwin, 2017). Consequently, city residents who managed to escape from excessive tourist traffic by avoiding historic city centers where tourism was focused, currently have no such option and have to live with many tourists who use accommodation services located directly in their vicinity (Ključnikov et al., 2018). Significant areas of conflict, therefore, emerge between residents and tourists including the shortage of available residential space, an increase in rents and the problem of constantly changing neighbors. As Gürsoy (2019) suggested, it is everyday life that is the place where tourismphobia becomes deeply rooted.

Methods

The aim of this study is to compare how overtourism is perceived by residents of large cities and important tourism resorts. In order to compare how the impact of overtourism on the lives of the residents was perceived, the research was carried out quantitatively using a questionnaire. Taking into account the complex nature of the overtourism phenomenon on residents, it was decided to ask the respondents a number of questions about how they perceived the impact of overtourism on their lives and the places they lived in. These questions were drawn from the literature review and were grouped into eight areas of influence (Number of tourists; Tourists' behavior; Prices and availability of goods and services, including housing; Economy and income; Overcrowding; Noise; Security; Destination development). Also, in each of the subject groups there were questions referring to both how overtourism affects the lives of individual residents and how the entire city or tourist resort operates.

The main part of the questionnaire contained requests to assess the impact of tourism on the everyday life of the destination's residents and entrepreneurs. Specifically, the respondents were asked about possible conflict situations in terms of excessive tourism. The respondents were asked to present their attitude towards specific views and opinions on the issues under consideration as well as to indicate whether they find certain views true or false. The answer variants were prioritized by assigning them subsequent ranks. The scale of grades used was a five-level character in the range from -2 to 2 (i.e. calibrated in 0 point) where values below zero mean objection to the opinion formulated while values above zero mean consent to the opinion. In addition, nine questions were focused on the general issues that were not closely related to each other or to other subject areas.

The research was conducted by the researchers themselves in the period between May and July 2020 in Poland. It was decided to conduct the research in two of the most important tourist destinations in Poland, a city (Cracow) and a tourist resort (Białka Tatrzańska). Cracow is the second largest city in Poland as well as the most crowded and the most frequently visited city destination in the whole country. The city was also the subject of several studies including those that analyzed the city in the context of overtourism (Plichta, 2019; Szromek, et al. 2020; Kruczek & Szromek, 2020; Zmyślony et al. 2020). Białka Tatrzańska is a new mountain resort based on high-standard modern ski infrastructure. It was converted from a forgotten pasture village to a fashionable ski resort within less than twenty years. It is also a heavily studied example in Poland (Mika, 2009; Krzesiwo, 2018) as it might be perceived as a typically catalytic resort (Szromek & Żemła, 2018) and most investments were made gradually by the local community, which truly benefited in terms of economy. Like many other places in Poland, Cracow and Białka Tatrzańska experienced a full lockdown at the time of the research which was gradually loosened during the survey period. This allowed the researchers to compare and analyze whether the lockdown-related disappearance of tourism from the places they lived in deprived the costs and benefits associated with tourism as well. Because of the rapidly changing situation, it was necessary to ask exact questions about specific moments of time. Therefore, the questions about overtourism were marked in the questionnaire as referring to the summer 2019 and winter 2019/2020 season, while the questions about lockdown were marked as referring to April 2020.

As many as 523 (371 in Cracow and 152 in Białka Tatrzańska) randomly selected respondents participated in the research. With the collected size of the sample, the assumed reliability threshold of the random sample were achieved (±5% maximum statistical error of the sample and p=0.95 confidence level were assumed). The questionnaire was analyzed for reliability using the Cronbach's alpha coefficient, which in this case was 0.8916. With this coefficient, we can be confident that the research tool was highly reliable.

The final stage of the research was the analysis of the obtained data. With average results and a comparative analysis using nonparametric statistical tests, deductions could have been made. The results were usually presented as the value of the arithmetic mean, supplemented by the standard deviation (xAv±SD), and sometimes also as the median value (Me) or percentage (%). The collected data were processed by performing a statistical analysis, both one-dimensional (in the form of a descriptive analysis) and two-dimensional (in the form of an analysis of the dependence of pairs of examined features). In some cases, the significance level of differences between the obtained mean values was also examined. The U (Mann-Whitney) test was used to compare two groups with variables with a distribution other than normal, while in the case of two samples of dependent variables, the Wilcoxon pair-order test was performed. The significance of differences between the structure indices was tested with the Chi-square test. To verify the statistical hypotheses, statistical tests were used taking into account the significance at the level of $\alpha \leq 0.05$. In some cases, statistical multivariate analyses could have been performed. Spearman's rank correlation coefficient was used to identify the relationship between the qualitative features.

Results

The statistical analysis of the data took into account questionnaires obtained from 523 respondents, including 70.9% (371) representatives of Cracow and 29.1% (152) from Białka Tatrzańska. There were more women (65.4%) than men (34.6%) among the respondents. Most of the respondents were people in the middle age of 30-50 (45.7%) and young age (34.8%). Mature people aged 50-70 accounted for 16.6% of the population in the study. Most of the respondents had higher education (67.5%) while 24.5% had only secondary education. The vast majority of the respondents are permanent residents of the towns (80.9%) while the others (19.1%) are people who lived outside the town, but carried out various professional activities in these places. The researchers collected data in various places of the surveyed destinations. Hence, 38.4% of the respondents lived in the center of the surveyed tourist destinations, 45.3% of the respondents lived around the center, (i.e. in districts adjacent to the city/village center) and the rest (16.3%) were based in city outskirts. The advantage of the sample is the fact that almost two out of three respondents (61.2%) are residents of tourist destinations who have been living there for over 20 years, and 13.4% have been living there for 11 to 20 years. Only 5.2% of the respondents had moved to this area in the last 2 years.

As it has already been mentioned, obtaining the respondents' opinions was based on a five-point scale containing extreme assessments of compliance with the previously formulated opinion. Then, these assessments were quantified, reducing them to a rating between -2 and 2, and then averaging the obtained scores. The close to zero (0.0 ± 0.05) result indicates that the respondents did not make a clear decision regarding subsequent opinions. On the other hand, an average score higher than 0.05 means that the overwhelming majority of respondents agreed with the formulated opinion, while the score lower than -0.05 means that the majority of respondents did not agree with the opinion. With two subsequent ranges, the ratings expressing consent or objection to the formulated opinions can be graded. The rating here that ranges $|0.05 \div 1.00|$ indicates moderate agreement or objection, while the rating in the $|1.01 \div 2.00|$ range expresses clear compliance or objection.

Taking into account the adopted methodology, a table of mean scores for individual opinions and a table of significance levels of differences between the average scores for individual subgroups studied were obtained. The results show that the responses of the residents of Cracow and Białka Tatrzańska differ significantly in some subject areas (Economy and income; Security; City development) and in some of the issues raised in the general part. However, in the case of other subject areas (Number of tourists and other questions in the general part), residents' views of both analyzed destinations are compliant. Finally, in the case of several subject areas (Tourists' behavior; Prices and availability of goods and services, including housing; Overcrowding), the answers were compliant on some issues, while in the case of other Białka Tatrzańska and Cracow residents' responses were different.

Both groups of respondents confirmed how burdensome the following consequences of overtourism were: too many tourists, non-compliance with the rules of culture by tourists, higher prices in shops and service outlets, and excessive noise. They also agreed to confirm that the result of an influx of tourists (i.e. the extension of the offer in cultural events) is an advantage. Moreover, the respondents from both groups did not notice that the large number of tourists who visit the city/village was a cause of their discomfort. Both groups' perceptions were also compliant about how they assessed the lockdown situation as the lack of tourists was one of the factors that resulted in a decrease in the city's/village's income and in the natural environment improvement. Moreover, in relation to the lack of tourists, residents stated that were more likely to stay in places attractive to tourists in the city/village arguing (to a moderate degree) that without tourists, the city/village is a place more friendly to live. Both groups of respondents also rejected the statement that they started to miss tourists (absent because of the pandemic). They were also compliant about having no clear opinion about the statement that a city/village without tourists loses part of its charm and whether they use catering establishments full of tourists.

In the case of the other answers, statistically significant differences between Białka Tatrzańska and Cracow residents' opinions can be seen. The Białka Tatrzańska residents more often complained about the impact of tourists' behavior on their lives including foreign tourists and the problems related to the impact of tourism with regard to safety. The Cracow residents do not really notice the problems with the overcrowding generated by tourists while for the Białka Tatrzańska residents, it is a big problem or even very big. On the other hand, the Cracow residents more strongly emphasized the disadvantages related to the large number of apartments rented to tourists in their vicinity and, as such, the availability of residential facilities for locals being reduced as their prices are rising. More than the Białka Tatrzańska residents, the Cracow residents do not notice that the quality of their life associated with the use of residential facilities for tourist rental was reduced or that noise impacts their everyday life. There is a very large discrepancy between the opinions of both groups on how they perceive economic benefits resulting from the development of tourism. While economic benefits are identified by the resort's residents, the city residents either do not experience them at all or only a small part of it. Similar comments can be applied to how the impact of tourism investments on the city's/village's development is perceived. The consequence of this perception is the suggestion of Białka Tatrzańska residents to limit their everyday life's comfort in favor of building a more attractive tourist offer, while the Cracow residents quite clearly reject this option. Nevertheless, it was also the Białka Tatrzańska residents who were less enthusiastic about the possibility of spending free time in their location and expressed greater hope for a positive transformation of tourism after the pandemic.

Even though the Cracow residents generally did not agree with the statement that tourist visits are one of the factors that resulted in their household income increase, there was a small group (43 people, 8.9% of respondents from this city) among them that fully or partially agree with this statement. Their opinions are worth paying attention to, because at the same time, as a separate group they moderately agree with the statement that they are inclined to limit their needs in favor of building a more attractive tourism offer (0.35 ± 1.21).

Discussion and conclusions

In light of increasing overtourism in both cities and tourist resorts, the development of new perspectives and tools to manage tourist destinations is needed. This mainly applies to the need to diversify the developmental approach for different destinations, and in particular, to take into account the peculiarity of cities as destinations. Special attention should be paid to the role that residents play in typical resort destinations and in cities. The results presented in the chapter are the first attempt to establish a new approach to tourism destinations management based on different attitudes towards destinations of different types.

The results of the study proved that there are significant differences between how tourist resort residents (Białka Tatrzańska) and large city (Cracow) residents perceive the consequences of intensive tourism development. Study results reveal that the negative effects of tourism development such as overcrowding, noise, or safety issues were more strongly raised by the Białka Tatrzańska residents, while new trends related to renting apartments for tourists (prices and real estate availability) were the subject of greater dissatisfaction for the Cracow residents.

The results also suggest that the most important reason for the different attitudes of the residents of cities and the residents of tourist resorts towards tourism development lies elsewhere. The greatest discrepancies between the representatives of both groups were noticed in how they assess the benefits of the tourism development. The Białka Tatrzańska residents experience quite a lot of them, both in their personal lives and in the development of the entire village while the Cracow residents do not experience such benefits in their lives at all and experience very little benefits in the city's development. As a result, it is the resort's residents who consider limiting their own needs in order to build a competitive offer of the destination and attract new tourists as reasonable. The Cracow residents, experiencing and expecting no such benefits, do not find it reasonable to limit their own needs in order to improve the offer for tourists. Therefore, it seems that since the current direction of research on the phenomenon of overtourism was strongly focused on the analysis of the negative consequences of the tourism development activities (Namberger et al., 2019, Phi, 2020; Szromek, et al., 2020), it should be urgently supplemented with an analysis of the benefits of the tourism development as it is perceived by the residents of particular destinations. In order to properly understand the essence of anti-tourism movements and tourism-phobia, the full statement of benefits and costs related to the tourism development should be made.

A closer analysis is also required into how the costs related to the tourism development are perceived. It is symptomatic that mainly the Cracow residents but also Białka Tatrzańska residents noticed the negative impact of tourism development on a quite abstract notion of a location, but they definitely less often noticed it in their own lives. It can, therefore, be assumed that the residents of destinations can isolate their private lives from the negative influences of tourism. By doing so, they only observe these influences but do not experience them. The study shows clearly significant differences between how the effects of tourism development are perceived by city residents and by the tourist resort residents. However, the fact that the research was conducted only in two destinations can be an essential limitation in making the results universal. The results obtained require future confirmation in the research that should be conducted in other destinations, especially in other cultural circles. It should also be kept in mind that the division of destinations into resorts and cities is blurry and does not fully reflect how rich the tourism market is. It does not take into account, among others, the status of medium-sized cities that have become almost entirely dependent on tourism in terms of economy such as Venice or Dubrovnik. They operate like tourist resorts and it is difficult to predict whether their inhabitants will show attitudes closer to those identified in Cracow or rather to those in Białka Tatrzańska.

The presented study illustrates specific attitudes of residents observed in a specific moment of exceptionally dynamic reality. At the time this chapter was written, there are still no unambiguous forecasts showing the expected directions of the tourism development during and after the pandemic threat. The scale of economic losses to be suffered by tourism and the entire hospitality industry in connection with the pandemic, how the travel directions will change, and, above all, how the number of tourists will change. These factors are also unknown.

Highlights

- ☐ The perception of the effects of overtourism varies between city residents and resort residents.
- ☐ City residents experience more negative effects of short-term residential rentals to tourists while resort residents are more likely to complain of overcrowding, noise, inappropriate tourist behaviour, and lower safety.
- ☐ The biggest difference is in how the benefits are perceived: city residents experience no benefits, while resort residents appreciate the economic effect of tourists' visits and they declare they can limit their own needs to make this effect as big as possible.

Questions

1. Why do city residents not experience the positive effects of tourism development?
2. Is it possible that the negative impact of tourism on the resort residents is so significant that they might also show anti-tourism attitudes despite economic benefits?
3. What may be the future direction of tourism development, especially of urban tourism once the threat of the COVID-19 pandemic is over?

References

Ashworth, G. J. (1989) Urban tourism: An imbalance in attention, in C. P. Cooper (ed.), *Progress in Tourism, Recreation and Hospitality Management*, London: Belhaven, pp. 33-54.

Ashworth, G. J. (1992) Is there an urban tourism?, *Tourism Recreation Research*, 17(2), 3–8.

Ashworth, G. J. (2003) Urban tourism: still an imbalance in attention, in C. Cooper (ed.) *Classic Reviews in Tourism*, Clevedon: Channel View, pp. 143-163.

Butler, R. W. (1980) The concept of a tourist area cycle of evolution: implications for management of resources, *Canadian Geographer*, 24(1), 5-12.

Canestrelli, E., & Costa, P. (1991) Tourist carrying capacity: A fuzzy approach, *Annals of Tourism Research*, 18(2), 295-311.

Capocchi, A., Vallone, C., Pierotti, M., & Amaduzzi, A. (2019) Overtourism: A literature review to assess implications and future perspectives, *Sustainability*, 11(12), 3303.

Christaller, W. (1964) Some considerations of tourism location in Europe: The peripheral regions-under-developed countries-recreation areas, *Papers of the Regional Science Association*, 12(1), 95-105.

DiNatale, S., Lewis, R., & Parker, R. (2018) Short-term rentals in small cities in Oregon: Impacts and regulations, *Land Use Policy*, *79*, 407-423.

Doxey, G. V. (1976) When enough's enough: The natives are restless in Old Niagara, *Heritage Canada*, 2(2), 26-27.

Goodwin, H. (2017) The challenge of overtourism, *Responsible Tourism Partnership Working Papers*, 4.

Gürsoy, İ. T. (2019) Beauty and the beast: A fairy tale of tourismphobia, *Tourism Planning & Development*, 16(4), 434 451.

Gürsoy, D., Jurowski, C., & Uysal, M. (2002) Resident attitudes: A structural modeling approach, *Annals of Tourism Research*, 29(1), 79-105.

Hall, C. M. (1994) *Tourism and Politics: Policy, power and place*. Oxford: John Wiley.

Herrera, L. M. G., Smith, N., & Vera, M. Á. M. (2007) Gentrification, displacement, and tourism in Santa Cruz de Tenerife, *Urban Geography*, 28(3), 276-298.

Horn, K., & Merante, M. (2017) Is home sharing driving up rents? Evidence from Airbnb in Boston, *Journal of Housing Economics*, 38, 14-24.

Ključnikov, A., Krajčík, V., & Vincúrová, Z. (2018) International sharing economy: The case of Airbnb in the Czech Republic, *Economics & Sociology*, 11(2), 126-137.

Koens, K., Postma, A., & Papp, B. (2018) Is overtourism overused? Understanding the impact of tourism in a city context, *Sustainability*, 10(12), 4384.

Kruczek, Z., and Szromek, A. R. (2020) The identification of values in business models of tourism enterprises in the context of the phenomenon of overtourism, *Sustainability*, 12(4), 1457.

Krzesiwo, K. (2018) Ocena wybranych działań promocyjnych ośrodków narciarskich w Polsce–przykład ośrodka narciarskiego Kotelnica Białczańska w Białce Tatrzańskiej, *Folia Turistica*, 48, 261-282.

Maitland, R. (2018) Introduction: National capitals and city tourism, in R. Maitland, & B. Richie, (eds.), *City Tourism: National Capital Perspectives*, Wallingford: CABI, pp. 1-18.

Mika, M. (2009) Ski tourism in the Polish Carpathians: present state and issues of development, *Folia Geographica*, 49, 198-208.

Milano, C., (2017) *Overtourism and Tourismphobia: Global trends and local contexts*. Barcelona: Ostelea School of Tourism & Hospitality.

Milano, C., Cheer, J. M., & Novelli, M. (2019) Introduction, in: Milano C., Cheer, J.M. & Novelli, M. (eds.), *Overtourism: Excesses, discontents and measures in travel and tourism*, Walligford: CABI, pp. 1-17.

Namberger, P., Jackisch, S., Schmude, J., & Karl, M. (2019) Overcrowding, overtourism and local level disturbance: how much can Munich handle?, *Tourism Planning & Development*, 16(4), 452-472.

Novy, J., & Colomb, C. (2016) Urban tourism and its discontents. An introduction, In: C. Colomb, & J., Novy (eds.), *Protest and Resistance in the Tourist City*. London: Routledge, pp. 1-30.

Peeters, P. M., Gössling, S., Klijs, J., Milano, C., Novelli, M., Dijkmans, C. H. S., & Mitas, O. (2018) *Research for TRAN Committee-Overtourism: impact and possible policy responses*. Brussels: European Parliament.

Phi, G. T. (2020) Framing overtourism: A critical news media analysis, *Current Issues in Tourism*, 23(17), 2093-2097.

Pizam, A. (1978) Tourism's impacts: The social costs to the destination community as perceived by its residents, *Journal of Travel Research*, 16(4), 8-12.

Plichta, J. (2019) The co-management and stakeholders theory as a useful approach to manage the problem of overtourism in historical cities–illustrated with an example of Krakow, *International Journal of Tourism Cities*, 4(5), 685-699.

Su, L., Huang, S., & Huang, J. (2018) Effects of destination social responsibility and tourism impacts on residents' support for tourism and perceived quality of life, *Journal of Hospitality & Tourism Research*, 42(7), 1039-1057.

Szromek, A. R., Kruczek, Z., & Walas, B. (2020) The attitude of tourist destination residents towards the effects of overtourism—Kraków sase study, *Sustainability*, 12(1), 228.

UNWTO (2018) *Overtourism? Understanding and Managing Urban Tourism Growth beyond Perceptions: Executive Summary*. Madrid: UNWTO.

Wegmann, J., & Jiao, J. (2017) Taming Airbnb: Toward guiding principles for local regulation of urban vacation rentals based on empirical results from five US cities, *Land Use Policy*, 69, 494-501.

Zmyślony, P., Kowalczyk-Anioł, J., & Dembińska, M. (2020) Deconstructing the overtourism-related social conflicts, *Sustainability*, 12(4), 1695.

8 Shedding light upon the shadows: Exploring the link between country destination image and dark tourism

Kladou Stella, Psimouli Maria, Skourtis Georgios, Giannopoulos Antonios and Eskitark Nilgün Handan

Abstract

Dark tourism is one of those tourism types that currently attracts growing interest and varying interpretations. The present study aims to determine the importance of specific dark tourism motives (namely those relevant to war and terrorism) for dark tourism development, taking destination and country image into consideration. The research was carried out in the context of Turkey, a developing country, which exhibits dark tourism potential in relation to the motives explored. An explanatory quantitative approach was followed. The results reveal the importance of motivations, previous experience, intention to visit and the significance of different tourism types according to destination image in a tourism context. The findings shed light on aspects that can drive dark tourism development, without falling into the fallacy of prioritizing a "better mousetrap (dark tourism site) that nobody will want".

Keywords:

Dark tourism, war, terrorism, culture, destination image, country image

Introduction

In the last decade and on the path to sustainable development, UNWTO (2012, 2021) has updated its definition of tourism types to embrace more sustainable and diversified types. Along these lines, dark tourism emerges as a niche market. It is connected to various definitions, subcategories of the topic and tourist motivations for visiting dark tourism sites (Mitchell et al., 2020). Villamediana-Pedrosa et al.'s study (2020) as well as the country destination image and tourism type match/mismatch framework (Kladou et al., 2014), highlights the importance of investigating travel motives together with the destination profile. Tourism is an image-driven industry (Lindblom et al., 2018) as destination images and beliefs affect choice and travel intention (De Nisco et al., 2015), and so are part of the overall destination branding strategy (Giannopoulos et al., 2020).

Among the various dark tourism places, sites associated with war (i.e. battlefields) comprise the largest category of tourist attractions globally (Smith, 1998) and their importance has been repeatedly emphasized (Chronis, 2005; Slade, 2003; Stone, 2012; Thompson, 2004). Next to this, commemorating victims of terrorism, and seeking to develop dark tourism primarily initiated from this motive, may not be adequate for all countries (Seraphin, 2017), highlighting the need to investigate the dark tourism potential in line with its match to destination image in a country context (Kladou et al., 2014).

Addressing this research gap, the present explanatory study focuses on the case of Turkey. The country is home to Gallipoli, one of the most popular dark tourism sites for international tourists. The site is particularly important for international travellers (mostly Australian and New Zealanders), who visited it in herds prior to the COVID-19 outbreak (Çakar, 2018; Slade, 2003), and is in close proximity to the putative site of Homer's Troy (Boz, 2020). Prior image studies (e.g. Tasci et al., 2007), the intensity of terorrism in Turkey and the publicity many attacks may, from time to time, see in international media suggest the possibility of commemorating victims of terrorism in a constructed site similar to that of Ground Zero in New York. An online questionnaire and subsequent analysis helped us frame the limitations of developing dark tourism initiatives when considering the importance of factors such as the country and destination image as well as potential visitors' cultural context and motivation.

Literature review

Exploring destination image in a country context

Both country and destination image are multidimensional concepts that reflect individuals' assessments of different features of places and embrace cognitive, affective, conative, or overall perceptions (Gallarza et al., 2002). The cognitive dimension mirrors people's knowledge and beliefs on a given place, whereas affective image includes emotions and feelings (Maher & Carter, 2011). The conative component then seeks to reflect the behavior, which may result from the individual's evaluation of the place (e.g., Choi et al., 2007). Studies have found that cognitive and affective country images are usually aligned, without though identifying whether one of the two antecedes the other, while both image dimensions seem to positively relate to destination beliefs (Lindblom et al., 2018; De Nisco et al., 2015). In sum, there are some common characteristics in the assessment of destination and country image. Thus, Alvarez and Campo (2014) as an alternative suggest summarizing the negative or positive cognitive and affective assessment into overall image.

Studies on country and destination image also reveal potential differences among the two, in terms of their definition and content. Country image is more generic and inclusive, whereas destination image refers specifically to those country associations that relate to the tourism perspective of the place (Mossberg & Kleppe, 2007). More importantly, individuals may hold a positive image for a given country as a destination, yet negatively comment on economic or political aspects (Alvarez & Campo, 2014). The latter can particularly be the case in developing countries (e.g., Campo & Alvarez, 2010, Alvarez & Campo 2014; Nadeau et al., 2008). Even on such occasions, however, Alvarez and Campo (2014) suggest that boosting tourism image can improve country image in relation to non-tourism characteristics (e.g., economic, political) as well.

On the other hand, Kladou et al. (2014), found that the perception of destination image in a country context tends to vary across different types of tourism. This means that specific primary travel motives affect willingness to visit a given country based on the match/mismatch of country destination image with the type of tourism. Thus, place stakeholders that decide to boost destination image in a country context, should explore the match/ mismatch framework and evaluate travel motives for the tourism type put in question.

In the pursuit of a competitive advantage, more niche forms of tourism have emerged. Dark tourism efforts materialize and mainly refer to visiting sites of death or associated with death (Foley & Lennon, 1996; Lennon & Foley, 2000). The term has interchangeably been described in the literature as 'thanatourism' (Slade, 2003; Seaton, 2002), while even more niche types relate to dark tourism such as heritage tourism (Biran & Hyde, 2013), battlefield tourism (Winter, 2011), secular pilgrimage (Hall et al, 2018; Hyde & Harman, 2011), suicide, atomic, grief, genocide, disaster, favela and conflict heritage tourism (Light, 2017).

Dark tourism experience may be performed in places that developed as dark tourism sites as history progressed (e.g., the Auschwitz concentration camp in Poland) or relate to sites, which have been explicitly developed to commemorate victims (e.g., Ground Zero in New York). In any case, most of the dark tourism types seem to have some connection with what UNWTO (2021) identifies as cultural tourism. On the other hand, some of the afore-mentioned niche types (such as battlefield, grief, disaster) reflect a thin line between dark and adventure tourism, taken that adventure tourism may *"involve some kind of real or perceived risk"* (UNWTO, 2021, p.36) and commemorating victims particularly of terrorism. What seems to distinguish between adventure and dark tourism is the image of the country and, more specifically, the perceived political risk (Liu & Pratt, 2017).

Travel motives: a focus on dark tourism

Dark tourism can help destinations in a country context to embrace the 'darker' sides of their past. Foley and Lennon (1996: 198) describe, dark tourism as "(...)*the phenomenon which encompasses the presentation and consumption (by visitors) of real and commodified death and disaster sites*", Stone (2006:146) refers to it as "*the act of travel to tourist sites associated with death, suffering or the seemingly macabre*", while Tarlow (2005) defines it as "*visitations to places where tragedies or historically noteworthy death has occurred and that continue to impact our lives*".

Seaton (1999) built on motivation in order to identify five categories of activities which relate to death tourism: 1) Travel to watch death (e.g. public executions); 2) Travel to sites after death has occurred (e.g. Auschwitz); 3) Travel to internment sites and memorials (i.e. graves and monuments); 4) Travel to (e.g. war) re–enactments; and 5) Travel to synthetic sites at which evidence of the dead has been assembled (i.e., relevant museums and sites). A decade later, reviewing the literature and adding up their research findings, Biran et al. (2011) classified travel motives associated with dark tourism.

With a focus on heritage, Tunbridge and Ashworth (1996) suggest atrocity can be a tourist attraction. Ideological and political aspects inherently embedded in most dark tourism sites call for the heritage parameter inclusion in the relevant studies (Seaton & Lennon, 2004). Within this framework, academics have linked the personal meaning assigned to a dark tourism site not only with one's motivation to visit it but also with one's experience (Sigala & Steriopoulos, 2021) and satisfaction with the visit (Poria et al., 2003; Poria et al., 2006).

Interestingly, many studies do not consider the variety of tourists' perspectives towards dark tourism sites that draw from the different levels of knowledge and familiarity they may have to the site explored (Waitt, 2000). Furthermore, in some cases visiting a dark site has been found to link to a feeling of moral obligation to the previous generations and the need to familiarize with tragic events and commemorate their victims (Thurnell-Read, 2009; Kang et al., 2012).

8

Before selecting a destination, tourists aim to match their expectations and motives with perceived destination aspects (e.g. previous experience, word of mouth, press reports) through an evaluation process (Buhalis, 2000; Kim et al., 2000). As a result, deep knowledge and understanding of travel motives has been underlined for the development of destination parameters that will enhance tourists' experience (Pansiri, 2014) in order to meet/exceed their expectations and thus strengthen the destination's competitiveness (Zouni & Kouremenos, 2008).

According to Stone (2013), dark tourism is an interdisciplinary field of study that has emerged through death education and tourism scholarship, while he also underlines the linkage between dark tourism and thanatopic tradition (Seaton, 1996) as well as the importance of different representations of mortality per culture (Howarth, 2007) for the field. Biran et al. (2011, p.832) claim that *"it is not death or the dead that should be considered, but living peoples' perception of them"*. Considering the socio-cultural context of death that differentiates not only its position and related values per culture (Mellor & Shilling, 1993) but also individuals' interpretation of death-related aspects (Kearl, 2009), Stone (2012) emphasizes that dark tourism can be considered as *"a cultural representation of death and dying"* (p. 1582). Within this framework, *"'darkness' is subjectively and socially constructed"* (Sigala & Steriopoulos, 2021, p.2) as the respective interpretations and experiences depend on one's socio-cultural background (Light, 2017). Experiential and emotional aspects of dark tourism (Oren et al., 2021) help in revealing its co-creative nature, with researchers arguing that dark tourism experiences are *"subjectively, phenomenologically and contextually determined based on the visitor's profile and situation"* (Sigala & Steriopoulos, 2021, p.2).

Many studies have attempted to profile tourists seeking dark tourism experiences to classify experiences accordingly (i.e. Dore, 2006; Raine, 2013). Commemorative dark tourism experiences were proposed as a specific type of dark tourism by Dore (2006) along with knowledge seeking, satisfying curiosity and cathartic experiences. Given the subjectivity embedded in dark tourism, people usually assign their own meanings to dark places and death that although they may reflect a different understanding than the one intended by the supplier (Du et al., 2013), increase the site's emotional impact on them (Cameron & Gatewood, 2003).

Dark tourism sites associated with war (i.e. battlefields) that constitute the largest category of tourism attractions at a global level (Smith, 1998) are considered to be of great importance (Thompson, 2004) due to their psychological and cultural aspects (Slade, 2003) that through symbolism link history to the social values of patriotism and national unity (Chronis, 2005; Stone, 2012). As a result, it comes as no surprise that respective motives point to nationhood (Hall et al., 2010). During the last decades, media have been highlighted as stimulating interest towards dark tourism sites (Kang et al., 2012) and creating related perceptions through the descriptions of the tragic events that took place there (Blom, 2000). Events and sites that have attracted media attention and coverage are usually of interest to dark tourists (Foley & Lennon, 1996; Stone, 2009). Consequently, sites associated with extraordinary disaster (i.e. terrorism and natural disasters) have been identified as a distinct type of dark tourism sites (Kang et al., 2012).

The case of Turkey

Turkey is a developing country showing elasticity in terms of how the country image may be assessed as positive to negative (Tasci et al., 2007; Baloglu & Mangaloglu, 2001; Campo & Alvarez, 2010). In fact, Turkey currently lies at the 56th position in the Good Country Index (Anholt, 2021). On the other hand, Turkey may experience negative image bias (e.g. Tasci et al., 2007) but is still a well-known destination, a country boasting of its history and heritage and positively evaluated as a destination. As Alvarez and Campo (2014) emphasized, problems such as terrorism, political and economic crises may lead to serious and permanent damage on destination image. Despite experiencing such economic and political issues for decades, Turkey is a country to which people would like to travel (Campo & Alvarez, 2010).

Representing one of the first dark tourism initiatives, the battlefield of Gallipoli has inspired the growth of a specific tourism product in the country, although it has not always been clearly identified as such in the literature (Çakar, 2018; Slade, 2003). Until the outbreak of the COVID-2019 pandemic, the site attracted many visitors who sought to commemorate ANZAC (Australian and New Zealand Army Corps) or were interested in the cultural history of Homer's Troy and, therefore, engaged funding to support the promotion of the site through various projects (e.g. Ministry of Culture and Tourism, 2021).

In recent years, Turkey is not the only country that has experienced terrorist attacks, although the country appears to have experienced more than those in other European countries combined. Most of these attacks relate to national political issues and did not seem to target international tourists in specific. Still, a deadly suicide bombing attack in January 2016 in the touristic center of Sultanahmet in Istanbul (e.g. The Guardian, 2016), sparkled discussions of whether to invest in international dark tourism by creating a site that, similar to Ground Zero in New York, would commemorate victims of terrorism. Such monuments and commemorations can be identified as a relatively 'popular' tactic in Turkey, especially when addressing domestic audiences (e.g. the well-known Bosporus Bridge is now 15 July Martyrs (*15 Temmuz Şehitler*) Bridge, named after the coup d' état attempted on 15 July 2016 against state institutions). Still, such an idea needs to be considered in relation to debates whether dark tourism should be considered as a type of tourism appropriate to develop regardless country and destination image or not (Seraphin, 2017).

Methods

The literature highlights the importance of dark tourism and specific relevant motives (i.e. related to war and terrorism), however, the importance of specific monuments (e.g. to commemorate victims of terorrism) or dark tourism initiatives to influence image perceptions and intention to visit remains unexplored. Investigating these aspects will help us shed light on the importance of a growing tourism type, namely dark tourism, taken the destination image in a country context match/mismatch framework (Kladou et al., 2014).

To address the research objective, this study adopted an explanatory, experimental approach. A structured questionnaire was used to examine the following:

1. Identify respondents' interest to travel driven by specific dark tourism motives, namely, to commemorate war veterans or victims of terrorism (Stone, 2012; Seraphin, 2017; Smith, 1998; Thompson, 2004).
2. Detect whether respondents have had some travel experience with Turkey and then measures respondents' perceptions on overall country and cognitive, affective and overall destination image (adjusted by Campo & Alvarez, 2010; Alvarez & Campo, 2014).
3. Explore whether after exposure to selected images, respondents were likely to visit Turkey driven by the dark motives in question (i.e. to commemorate war victims and to commemorate victims of terrorism). At this point, respondents were shown photographs of Gallipoli and sites where terrorism victims could be commemorated. Photographs remind us of inevitable mortality (Barthes, 2000), and are also important for destination image (Deng & Li, 2018).

A 7-point Likert scale was applied for all questions but the demographic ones. The survey was conducted online in spring 2019, through a Facebook ad that invited people to respond based on a set profile and criteria (i.e., mostly interest in travelling and particularly in dark tourism contexts). A prior pilot study with international students had indicated that the questionnaire and selected photographs were adequate. A total of 193 fully completed questionnaires were retained, while reliability was achieved (Cronbach's alpha values > 0.85) for both country image and destination constructs. Following the margin of error formula, the sampling error in our study is estimated at 7.05%.

Findings

Before proceeding with the analysis, understanding the respondents' profile is necessary. Analysis illustrates that out of the 193 people who participated in the study, 62.7% had previous travel experience in Turkey. It should also be highlighted that 37.4% have visited Turkey up to 10 times. The respondents comprised of 42.5% men, 53.9% women and 3.6%, who preferred not to say. Almost half of them (45.6%) were up to 30 years old, 26.9% were 31-40 years old and the remaining 27.4% aged above 40 years old. The purpose of their visit was leisure (43%), business (11.4%) and VFR (24.9%) among others. Regarding their educational background, 19.7% have graduated from elementary and high school, 29.1% have completed undergraduate studies, 26% hold a master's degree and 7.1% hold a PhD.

People from different countries participated in the survey, currently residing in Germany, USA, UK, Spain, France, Poland, Australia, Ireland, Italy, the Netherlands, Belgium, Hungary, Greece, Norway, Russia, Slovakia, Algeria, Argentina, Brazil, Colombia, Cyprus, Denmark, Estonia, Finland, Hong Kong, India, Japan, Peru, Romania, Saudi Arabia, Serbia, Singapore, Slovenia, Taiwan and Uzbekistan. Regarding the nationality of the respondents, the largest groups came from the UK (25.4%), the USA (15%) and Germany (14.5%), while 33.7% came from other European countries, 4.7% from Asia and the Pacific region (including Australia, New Zealand, Japan), 3.6% from other American countries (e.g. Mexico, Peru, Brazil) and 3.6% from Africa and the Middle East (including Algeria, Saudi Arabia, South Africa).

An exploratory factor analysis (EFA) was performed to determine the underlying dimensionality of DI. The Kaiser–Meyer–Olkin (KMO) measure of sampling adequacy equaled 0.907 and Bartlett's test was statistically significant at $p < 0.000$ at 1544,572. The results of principal component factor analysis with varimax rotation showed a two-factor solution representing 66.59% of the total variance. The Cronbach's alpha reliability for the 2 dimensions varied from 0.85 to 0.87, indicating internal consistency.

The relationship between overall country image (as measured by the positive to negative) and visiting intentions (as measured by the visitation for war (VWV) and for terrorism veterans (VTV)) was investigated using

Pearson correlation coefficient. There was a weak, negative correlation between CI and visitation for war veterans VWV, r =-.22, n=193, $p < .0005$ with high levels of negative CI associated with lower visiting intentions for VWV. There was also a weak, negative correlation between CI and visitation for terrorism veterans VTV, r =-.21, n=193, $p < .0005$ with high levels of negative CI associated with lower visiting intentions for VTV.

In the same vein, the relationship between overall destination image DI (as measured by the positive to negative) and visiting intentions (as measured by the visitation for war and for terrorism veterans) was investigated using Pearson correlation coefficient. There was a strong, negative correlation between DI and VWV, r =-.17, n=193, $p < .05$ with high levels of negative DI associated with lower visiting intentions for VWV. There was no statistically significance correlation between DI and VTV, r =-.41,n=193, $p < .057$ (see the table 1 below).

Table 8.1: Correlations

Correlations (N=193)

		visitation for war veterans	visitation for terrorism veterans	positive to negative
visitation for war veterans	Pearson Correlation	1	0.831**	-0.173*
	Sig. (2-tailed)		0.000	0.016
visitation for terrorism veterans	Pearson Correlation	0.831**	1	-0.041
	Sig. (2-tailed)	0.000		0.573
positive to negative	Pearson Correlation	-0.173*	-0.041	1
	Sig. (2-tailed)	0.016	0.573	

**. Correlation is significant at the 0.01 level (2-tailed).

*. Correlation is significant at the 0.05 level (2-tailed).

A series of ANOVA were also conducted for the purpose of this study.

A one-way between-groups analysis of variance was conducted to explore the impact of nationality on motivation to commemorate war veterans, as measured by the MCWV. Participants were divided into 7 groups according to their nationality (1: Other American Countries; 2: Other European Countries; 3: UK; 4: USA; 5: Germany; 6: Asia and the Pacific; 7: Africa and Middle East). There was a statistically significant difference at the $p < .05$ level in MCWV scores for the two nationality groups: F (6, 186) = 2.2, p = .03. Post-hoc comparisons using the Tukey HSD test unveiled a lower motive to commemorate war veterans for other European countries (2.72) than Africa and Middle East (4.83).

In the same vein, a one-way between-groups analysis of variance was conducted to explore the impact of nationality on motivation to commemorate terrorism veterans, as measured by the MCTV. Participants were divided into 7 groups according to their nationality (1: Other American Countries; 2: Other European Countries; 3: UK; 4: USA; 5: Germany; 6: Asia and the Pacific; 7: Africa and Middle East). There was a statistically significant difference at the $p < .05$ level in MCTV scores for the five nationality groups: $F(6, 186) = 2.9$, $p = .01$. Post-hoc comparisons using the Tukey HSD test indicated higher motive to commemorate terrorism veterans for Africa and Middle East (5.0) than USA (2.66). UK (2.63), Germany (2.96), and other European Countries (2.60).

Last, a one-way between-groups analysis of variance was conducted to explore the impact of country of residence on visiting intentions for terrorism veterans, as measured by the VITV. Participants were divided into 4 groups according to their country of residence (1: UK; 2: USA; 3: Germany; 4: Rest of the world). There was a statistically significant difference at the $p < .05$ level in VITV scores for the two country of residence groups: $F(3, 189) = 2.8$, $p = .03$. Post-hoc comparisons using the Tukey HSD test indicated that visiting intentions to commemorate terrorism veterans in Turkey are lower for UK residents (2.27) than German residents (3.31).

Finally, independent sample t-tests were used to evaluate the effects of CI on previous experience in visiting Turkey. A statistically significance difference ($p < 0.1$) showed that the group who has no previous experience in Turkey (M=3.26, SD=1,8) has the more negative image compared to those who have previous experience (M=2.7, SD=1.6)

Discussion

This study investigates the role of specific dark tourism motives in relation to the country and destination image of Turkey. According to the findings, exhibiting the dark tourism potential of Turkey and a monument built to commemorate terrorism victims as an initiative to support dark tourism does not actually influence intention to visit the country. Adapting du Cros and McKercher's (2015) typology, this suggests that dark tourism may be important for purposeful dark tourists, who thereby seek a deep dark tourism experience (Oren et al., 2021) and have decided to visit the specific destination primarily motivated by the importance of dark tourism. On the other hand, efforts to develop dark tourism do not appear to matter for those who are sightseeing, casual or incidental dark tourists. Exploring the role of nationalities seems to reinforce the significance of this findings, since the socio-cultural background (Light, 2017) exhibits lower or higher motivation to visit Turkey urged by a specific dark tourism motive for some nationalities and socio-cultural identities compared to others.

Furthermore, the research examines the dark tourism potential in relation to destination image in a country context. In line with Kladou et al. (2014), country and destination image appear relevant for the development of (dark) tourism initiatives. Findings illustrate that experience with Turkey shifts the country image perceptions to more positive evaluations. On the other hand, there is no such shift on destination image, since (potential) tourists already positively evaluate Turkey as a destination. In relation to dark tourism, commemorating war veterans through dark tourism appears of significance, whereas dark tourism with a focus on terorrism aspects does not exhibit a similar potential. These findings further emphasize the difference between destination and country image for the country in question (as in Campo & Alvarez, 2010) and the importance attributed to such aspects prior to the development of such 'sensitive' tourism initiatives (Seraphin, 2017).

Limitations of the present research pave the way for future studies and can be summarized in the following points. First, the selection of only one country (Turkey) as a potential dark tourism destination shapes a very specific context of research, especially given the importance of culture for dark tourism experiences. Historical ties between Turkey and the respondents' home countries could affect answers regarding dark tourism, country and destination images. Thus, a cross-cultural research approach could result into more generalizable findings. Second, it would be interesting to check whether examination of types of dark tourism sites not relating to war and terrorism provide different results for the same country-respondents' combination used hereby. Last, studies exploring the potential impact of demonstrating different types of photos (even of the same site) could reveal whether there is a linkage between the depiction of a dark tourism site (selection of site to be depicted and way of depiction) and the respective associations that come to respondents' minds in terms of country/destination image and visit intention.

Conclusions

Drawing from the literature review and in combination with the findings of the present research, it seems that countries aiming for the development of a dark tourism element should first consider potential target audiences and motives (Villamediana-Pedrosa et al., 2020). Dark tourism motives and experiences rely on individuals' cultural context, thus visitors' nationality is inextricably linked not only to their intention to visit a dark tourism destination but also to their emotional engagement and satisfaction from the respective experience as well. Moreover, war-related dark tourism sites seem to be of a higher importance compared to terrorism-related ones, and are clearly associated with history and the historical ties among countries, thus perhaps knowledgeable audiences could more easily be attracted to such sites. On the other hand, terrorism-related dark tourism sites are mainly communicated and described through the media and this is the main factor to stimulate visitors' interest. As a result, media-related activity could transform tragic terrorist attacks to dark tourism commemorative sites, but even such efforts to prioritize dark tourism are subject to destination image in a country context and tourism type match/mismatch (Kladou et al., 2014).

The present study was conducted prior to the COVID-19 pandemic that has resulted in an unprecedented global situation dominated by the fear of death. Respective socio-cultural attitudes have been affected, thus willingness to travel in general and to visit dark tourism sites in specific needs to be reconsidered within this new framework. The pandemic could be considered as a 'dark' period in the human history and, as such, opens new avenues for research related to mortality, death, and its inclusion (or not) in the contemporary societies around the world.

Highlights

- Dark tourism may be important for purposeful dark tourists, who thereby seek a deep dark tourism experience and decide to visit a destination driven by the importance of dark tourism motives.
- Dark tourism initiatives are not significant for sightseeing, casual or incidental dark tourists.

- ☐ Nationality, as it is connected to national and cultural identities plays a significant role in identifying the importance of specific dark tourism motives.
- ☐ Dark tourism should develop with a focus on specific initiatives that match the country image to influence relevant visit intentions.
- ☐ Visiting experience with the destination leads to more positive country evaluations.

Questions

1. Have you ever visited a site/ monument 'commemorating death' (e.g. as a high school student)? What was the site commemorating?
2. Name a dark tourism site you are interested in and the country it is located in. Would you like to visit this site and country and why?
3. Assume that it is possible to visit a monument commemorating terrorism victims in the USA, France and Israel as part of a field trip organized by your university. Your teachers ask you to vote the destination of your choice. Which site and country do you choose? Identify and discuss three factors that drive your choice.

References

Alvarez, M. D., & Campo, S. (2014) The influence of political conflicts on country image and intention to visit: A study of Israel's image, *Tourism Management*, 40, 70-78.

Anholt (2021) The Good Country Index, available at https://index.goodcountry.org/ (accessed on 18 January 2021).

Baloglu, S. & Mangaloglu, M. (2001). Tourism destination images of Turkey, Egypt, Greece, and Italy as perceived by US-based tour operators and travel agents. *Tourism Management*, 22, 1 -9.

Barthes, R. (2000 [1980]) *Camera Lucida* (Trans. R. Howard), London: Vintage.

Biran, A., & Hyde, K. F. (2013) New perspectives on dark tourism, *International Journal of Culture, Tourism, and Hospitality Research*, 7(3), 191–198.

Biran, A., Poria, Y. & Oren, G. (2011) Sought experiences at (dark) heritage sites, *Annals of Tourism Research*, 38(3), 820-841.

Blom, T. (2000) Morbid tourism: a postmodern market niche with an example from Althorp, *Norwegian Journal of Geography*, 54 (1), 29-36.

Boz, M. (2020) Myths and legends in destination tourism marketing: the story of Hero and Leander—Canakkale, Turkey', in Coşkun İ., Lew A., Othman

N., Yüksek G., & Aktaş S. (eds), *Heritage Tourism Beyond Borders and Civilizations*, Singapore: Springer, pp. 3-14.

Buhalis, D. (2000) Marketing the competitive destination of the future. *Tourism Management*, 21(1), 97–116.

Çakar, K. (2018). Experiences of visitors to Gallipoli, a nostalgia-themed dark tourism destination: An insight from TripAdvisor. *International Journal of Tourism Cities*, 4(1), 98-109.

Cameron, C. M., & Gatewood, J. B. (2003) Seeking numinous experiences in the unremembered past, *Ethnology*, 42 (1), 55–71.

Campo, S. & Alvarez, M. (2010) Country versus destinaiton image in a developing country, *Journal of Travel & Tourism Marketing*, 27 (7), 748-764.

Choi, S., Lehto, X.Y., & Morrison, A. M. (2007) Destination image representation on the web: Content analysis of Macau travel related websites, *Tourism Management*, 28 (1), 118-129.

Chronis, A. (2005) Co-constructing heritage at the Gettysburg Storyscape, *Annals of Tourism Research*, 32 (2), 386–406.

Deng, N. & Li, X. (2018) Feeling a destination through the 'right' photos: A machine learning model for DMOs' photo selection, *Tourism Management*, 65, 267-278.

De Nisco, A., Mainolfi, G., Marino, V., & Napolitano, M. (2015) Tourism satisfaction effect on general country image, destination image, and post-visit intentions. *Journal of Vacation Marketing*, 21(4), 305-317

Dore, L. (2006) Gallipoli, a visitor profile. *Historic Environment*, 19(2), 46–51.

Du, W., Littlejohn, D., & Lennon, J. (2013) Place identity or place identities: The memorial to the victims of the Nanjing massacre, China, in White, L. & Frew, E. (eds.), *Dark Tourism and Place Identity: Managing and interpreting dark places*, Routledge, pp. 46–59.

Du Cros, H. & McKercher, B. (2014) *Cultural Tourism*, Routledge.

Foley, M., & Lennon, J. J. (1996) JFK and dark tourism: A fascination with assassination, *International Journal of Heritage Studies*, 2(4), 198–211.

Gallarza, M.G., Saura, I.G. & Calderón García, H. (2002) Destination image: Towards a conceptual framework, *Annals of Tourism Research*, 29 (1), 56-78.

Giannopoulos, A., Piha,L., & Skourtis, G. (2020) Destination branding and co-creation: a service ecosystem perspective, *Journal of Product & Brand Management*, (in press).

Golden, D. (1996) The museum of Jewish Diaspora tells a story, in Selwyn, T. (ed.), *The Tourist Image, Myth and Making in Tourism*. New York: John Wiley, 223–250.

Hall, J., Basarin, V. J., & Lockstone-Binney, L. (2010) An empirical analysis of attendance at a commemorative event: ANZAC day at Gallipoli, *International Journal of Hospitality Management*, 29 (2), 245–253.

Hall, J., Basarin, V. J., Lockstone-Binney, L., Yusuf, A., Winter, C., & Valos, M. (2018) Spiritual values and motives of secular pilgrims, *International Journal of Consumer Studies*, 42, 715–723.

Hyde, K. F., & Harman, S. (2011) Motives for a secular pilgrimage to the Gallipoli battlefields, *Tourism Management*, 32(6), 1343–1351.

Howarth, G. (2007) *Death and Dying: A Sociological Introduction*, Cambridge: Polity Press.

Kang, E-J., Scott, N., Lee, T.J., & Ballantyne, R. (2012) Benefits of visiting a 'dark tourism' site: The case of the Jeju April 3rd Peace Park, Korea, *Tourism Management*, 33, 257-265.

Kearl, M. (2009) *Kearl's Guide to Sociological Thanatology: Sociology of death and dying*, available at http://www.trinity.edu/mkearl/death.html (accessed 18 January, 2021).

Kim, S.-S., Crompton, J. L., & Botha, C. (2000) Responding to competition: A strategy for Sun/Lost City, South Africa, *Tourism Management*, 21(1), 33–41.

Kladou, S., Giannopoulos, A. & Assiouras, G. (2014) Matching tourism type and destination image perceptions in a country context, *Journal of Place Management and Development*, 7 (2), 141-152.

Lennon, J.J. & Foley, M. (2000) *Dark Tourism: The attraction of death and disaster*, Thompson Learning.

Light, D. (2017) Progress in dark tourism and thanatourism research: An uneasy relationship with heritage tourism, *Tourism Management*, 61, 275–301.

Lindblom, A., Lindblom, T., Lehtonen, M.J., & Wechtler, H., (2018) A study on country images, destination beliefs, and travel intentions: A structural equation model approach, *International Journal of Tourism Research*, 20, 1-10.

Liu, A., & Pratt, S. (2017). Tourism's vulnerability and resilience to terrorism. *Tourism Management*, 60, 404-417.

Maher, A.A. & Carter, L.L. (2011) The affective and cognitive components of country image: Perceptions of American products in Kuwait, *International Marketing Review*, 28 (6), 559-580.

Mellor, P., & Shilling, C. (1993) Modernity, self-identity and the sequestration of death, *Sociology*, 27, 411–431.

Ministry of Culture and Tourism (2021) Çanakkale *Historical Peninsula of Gallipoli*, available at https://www.ktb.gov.tr/EN-114066/canakkale-hisctorical-peninsula-of-gallipoli.html (accessed 19 January, 2021).

Mitchell, V., Henthorne, T.L. & George, B. (2020) Making sense of dark tourism: typologies, motivations and future development of theory', in Korstanje, M.E. & Seraphin, H. (ed.) *Tourism, Terrorism and Security (Tourism Security-Safety and Post Conflict Destinations)*, Emerald Publishing, pp. 103-114.

Mossberg, L. & Kleppe, I.A. (2007) Country and destination image – different or similar concepts? *Service Industries Journal*, 25 (4), 493-503.

Nadeau, J., Heslop, L., O'Reilly, N. & Luk, P. (2008) Destination in a country image context, *Annals of Tourism Research*, 35 (1), 84-106.

Oren, G., Shani, A., & Poria, Y. (2021) Dialectical emotions in a dark heritage site: A study at the Auschwitz death camp, *Tourism Management*, 82, In Press.

Pansiri, J. (2014) Tourist Motives and Destination Competitiveness: A Gap Analysis Perspective. *International Journal of Hospitality and Tourism Administration*, 15, 217-247.

Poria, Y., Butler, R., & Airey, D. (2003) The core of heritage tourism: Distinguishing heritage tourists from tourists in heritage places, *Annals of Tourism Research*, 30(1), 238–254.

Poria, Y., Reichel, A., & Biran, A. (2006) Heritage site management: Motivations and expectations, *Annals of Tourism Research*, 33(1), 1172–1188.

Raine, R. (2013) A dark tourist spectrum. *International Journal of Culture, Tourism and Hospitality Research*, 7(3), 242–256.

Seaton, A. (1996) Guided by the dark: from thanatopsis to thanatourism. *International Journal of Heritage Studies*, 2(4), 234-244.

Seaton, A. V. (1999). War and thanatourism: Waterloo 1815–1914. *Annals of Tourism Research*, 26(1), 130-158.

Seaton, A. V. (2002) Thanatourism's final frontiers? Visits to cemeteries, churchyards and funerary sites as sacred and secular pilgrimage, *Tourism Recreation Research*, 27(2), 73–82.

Seaton, A. V., & Lennon, J. J. (2004) Thanatourism in the early 21st century. Moral panics, ulterior motives and alterior desires, in Singh T. V. (ed.), *New Horizons in Tourism: Strange experiences and stranger practices*, Wallingford: Cabi, pp. 63–82.

Seraphin, H. (2017) Terrorism and tourism in France: The limitations of dark tourism, *Worldwide Hospitality and Tourism Themes*, 9(2), 187-195.

Sigala, M., & Steriopoulos, E. (2021). Does emotional engagement matter in dark tourism? Implications drawn from a reflective approach. *Journal of Heritage Tourism*, 1-21.

Slade, P. (2003) Gallipoli thanatourism: The meaning of ANZAC, *Annals of Tourism Research*, 30(4), 779–794.

Smith, V. (1998) War and tourism: An American ethnography, *Annals of Tourism Research*, 25(1), 202–227.

Stone, P. (2013) Dark tourism scholarship: a critical review, *International Journal of Culture, Tourism and Hospitality Research*, 7(3), 307-318.

Stone, P.R. (2012) Dark tourism and significant other death: Towards a model of mortality mediation, *Annals of Tourism Research*, 39(3), 1565-1587.

Stone, P. R. (2009) Dark tourism: morality and new moral spaces, in Sharpley, R. & Stone P.R. (eds.), *The Darker Side of Travel*, Bristol: Channel View, pp. 56-72.

8

Stone, P. R. (2006) A dark tourism spectrum: Towards a typology of death and macabre related tourist sites, attractions and exhibitions, *Tourism: An Interdisciplinary International Journal*, 54(2), 145–160.

Tarlow, P.E. (2005) Dark tourism: the appealing 'dark' side of tourism and more. In M.Novelli (ed) *Niche Tourism – Contemporary issues, trends and cases.* Oxford: Elsevier Butterworth-Heinemann: 47-58.

Tasci, A.D.A., Gartner, W.C. & Cavusgil, S.T. (2007) Measurement of destination brand bias using a quasi-experimental design, *Tourism Management*, 28, 1529-1540.

The Guardian (2016) *Deadly Istanbul blast 'caused by Isis suicide bomber'*, available at https://www.theguardian.com/world/2016/jan/12/explosion-istanbul-square-sultanahmet-turkey (accessed 19 January 2021).

Thompson, C. (2004) *The 25 Best World War II Sites: European Theatre*. San Francisco: Greenline Publications.

Thurnell-Read, T. P. (2009) Engaging Auschwitz: an analysis of young travellers' experience of Holocaust tourism, *Journal of Tourism Consumption and Practice*, 1(1), 26-52.

Tunbridge, J.E. & Ashworth, G.L. (1996) *Dissonant Heritage: The management of the past as a resource in conflict*. John Wiley & Sons: Chichester, UK.

UNWTO (2012) *Tourism Highlights 2012*, available at http://mkt.unwto.org/en/publication/ unwto-tourism-highlights-2012-edition (accessed 12 November 2012).

UNWTO (2021) *Tourism Definitions*, available at https://www.e-unwto.org/doi/epdf/10.18111/9789284420858 (accessed 18 January, 2021).

Villamediana-Pedrosa, J.D, Vila-Lopez, N. & Küster-Boluda, I. (2020) Predictors of tourist engagement: Travel motives and tourism destination profile, *Journal of Destination Marketing and Management*, 16, (in press).

Waitt, G. (2000) Consuming heritage: Perceived historical authenticity, *Annals of Tourism Research*, 27(4), 835–862.

Winter, C. (2011) Battlefield visitor motivations: Explorations in the great war town of Ieper, Belgium, *International Journal of Tourism Research*, 13(2), 164–176.

Zouni, G., & Kouremenos, A. (2008) Do tourism providers know their visitors? An investigation of tourism experience at a destination, *Tourism and Hospitality Research*, 8(4), 282–297.

9 Tourism suppliers' view of the role of government initiatives and tourism competitiveness in distressed contexts

Erisher Woyo

Abstract

This chapter analyses the role of government initiatives in tourism competitiveness, using data collected from suppliers operating in a distressed tourist destination. The government's involvement in the process of tourism development, especially in developing countries, is critical. The role and participation of governments in tourism development vary from minimal to a high level of involvement. Developing economics with ongoing political and economic challenges like Zimbabwe have not been comprehensively researched, especially on the nexus between competitiveness and government initiatives. Using qualitative data from a convenience sample of 15 hospitality and tourism managers in Zimbabwe, it was found that the role of government is important for enhancing tourism competitiveness. The study concludes that a higher level of government involvement is needed for Zimbabwe to enhance competitiveness. The government should play a greater role, especially in providing an enabling environment for improved competitiveness while reducing corruption.

Keywords:

Government initiatives; distressed destination; tourism competitiveness; Zimbabwe

Introduction

Before the COVID-19 pandemic, international tourism was projected to reach 1.8 billion arrivals, and emerging economies were expected to have 57% of the share (UNWTO, 2019). Though tourism growth has been revised downward due to the challenges emanating from the COVID-19 pandemic, tourism still plays a critical role in the economic development of emerging economies, especially job creation and revenue generation (Lee, 2015; Woyo & Slabbert, 2020). Several destinations globally are increasing government support to promote tourism development based on the sector's contribution to gross domestic product (GDP), especially among developing economies (Crouch, 2011; Woyo, 2018). Webster and Ivanov (2014) argue that the increased spending of public resources is justified, based on the need to attract more tourist arrivals and enhance destination competitiveness.

Competitiveness is critical for a tourist destination to globally obtain a favourable market position (Leung & Baloglu, 2013; Woyo & Slabbert, 2021). Pike and Page (2014) argue that managing competitive tourist destinations is critical in studying the tourism industry. Though destination competitiveness has been an active area of research, there is a need for comprehensive research to generate insights that destinations can apply in developing strategic plans (Cronjé & du Plessis, 2020). Past studies show a shifting trend in terms of how research is investigating destination competitiveness from simply focusing on aspects of competitive advantage to emphasising the role of private and public actors in the development equation (Armenski et al., 2018; Hall & Campos, 2014; Mei et al., 2015). This implies that even destinations with ongoing crises like Zimbabwe (Woyo & Slabbert, 2020; Woyo, 2021) need to understand what competitiveness means, its models and factors (Cronjé & du Plessis, 2020) and the role the government plays in enhancing competitiveness.

Government initiatives are imperative for enhancing destination competitiveness (Croes, 2011; Croes & Kubickova, 2013). However, studies investigating the role of government in a tourism competitiveness context (see Table 9.1) are relatively fewer (Javed & Tučková, 2020), suggesting an area of research that is still in its infancy (Hall & Campos, 2014; Wan & Bramwell, 2015). This is regardless of three decades of tourism competitiveness research (Kubickova & Martin, 2020). Though the role of government in tourism development among developing countries is prominent (Yang et al., 2008), the challenge is that research is yet to provide a nuanced understanding of the effectiveness of government initiatives and policies in enhancing tourism competitiveness (Yang et al., 2008; Kubickova, 2019), specifically from a destination with ongoing political and economic challenges like Zimbabwe (Woyo & Slabbert, 2020). This demonstrates the complexity of the relationship between government role and tourism competitiveness.

A review of past research shows that most studies on competitiveness focused more on definitional aspects, its measurements, and determinants (Abreu-Novais et al., 2016; Cronjé & du Plessis, 2020). Furthermore, African destinations, particularly those with ongoing political and economic challenges, have not been investigated (see Table 9.1). The focus of existing studies has been more on developing countries located in the Central American region. Moreover, a population gap is observed in the literature. Most of the studies were conducted using secondary data, drawn mainly from the World Trade Organisation, World Bank, Central Banks, Tourism Boards and the World Travel and Tourism Council. Therefore, methodologically, this study contributes to insights drawn from the tourism suppliers' primary data. Table 9.1 also shows that the same research gap was presented. This chapter responds to the need to understand how government initiatives and competitiveness are linked (Kubickova & Martin, 2020). Acknowledging these research gaps, this chapter investigates tourism establishments managers' perceptions regarding the government's role in enhancing destination competitiveness in a distressed context. This general aim is divided into two specific objectives: (1) identifying the government initiatives, (2) and their perceived influence on destination competitiveness in a distressed destination.

This study contributes to the literature by discussing the role of government initiatives using a destination in distress. Thus, the study's findings provide an interesting perception of government initiatives' impact on destination competitiveness. Understanding government initiatives' influence in enhancing a destination's competitiveness within a fluid political environment could help the supply side plan and develop the tourism industry better. Methodologically, this chapter provides findings based on primary data that has been qualitatively analysed in a field where most publications have been based on secondary data analysed quantitatively (see Table 9.1).

Table 9.1: Overview of research on government role and tourism competitiveness

Source	Study focus	Geographic focus	Methodology	Study population
Kubickova & Martin (2020)	Relationship between government and destination competitiveness	58 countries from developed and developing countries	Conceptual	Secondary data
Kubickova & Li (2017)	Tourism competitiveness, government, and tourism area life cycle	Costa Rica, Guatemala, and Honduras	A case study using time series analysis	Secondary data
Javed & Tučková, 2020	Government role in tourism competitiveness and its relationship with tourism area life cycle	Asia Pacific countries: China, India, and Pakistan	An econometric methodology using annual time-series data	Secondary data was used
Croes & Kubickova, 2013	Performance-based tourism competitiveness index	Belize, Costa Rica, El Salvador, Guatemala, Honduras, Nicaragua, and Panama		Secondary data
Kubickova (2016)	Role of government in tourism, linking competitiveness, freedom and developing economies	Belize, Costa Rica, El Salvador, Guatemala, Honduras, Nicaragua, and Panama	An econometric methodology using panel data analysis	Secondary data
Kubickova (2019)	Impact of government policies on destination competitiveness	Belize, Costa Rica, El Salvador, Guatemala, Honduras, Nicaragua, and Panama	A quantitative methodology using regression analyses	Panel data from the 7 Central American countries, ranging from 1995 to 2013, for a total of 1134 observations

Literature review

Tourism competitiveness

Competitiveness is complex and difficult to measure in a tourism context (Dodds & Holmes, 2020; Woyo & Slabbert, 2021). This complexity is also reflected in Table 9.1 where the research gap appears to be the same and not solved across a range of studies. There is no universally accepted definition of the term (Dodds & Holmes, 2020; Woyo, 2018) and measuring constructs due to differences in destinations and countries. However, most definitions focused mainly on the destination's ability to increase tourism spending, attract more visitors, provide memorable experiences, optimise destination attractiveness, deliver appealing tourism offerings, increase market share, and enhance the well-being of residents sustainably (Dodds & Holmes, 2020; Ritchie & Crouch, 1999). This makes the process of defining and measuring tourism competitiveness a tenuous activity (Kubickova & Martin, 2020).

The development of new tourism products, markets and customers requires destinations to think more like businesses (Kubickova & Martin, 2020). This is critical for creating and sustaining competitive advantage (Cronjé & du Plessis, 2020; Kubickova & Martin, 2020: Woyo & Slabbert, 2021). It has been argued that countries and regions which rely on tourism income must create a competitive advantage to maintain their market position (Gooroochurn & Sugiyarto, 2005). However, creating destination competitive advantage is intricate due to the complexity of the field and the involvement of multiple private and public sector stakeholders (Javed & Tučková, 2020; Tučková & Jurigová, 2014).

Government involvement and competitiveness

Government involvement in tourism development is critical for economic performance and modern economic activities (Glaeser & Shleifer, 2002). For tourism destinations to be successful, there is a need for a "*whole-of-government approach to tourism development and management*" (World Travel Council, 2015:30). Hall (2007:219) notes that this level of tourism performance is achieved when the government establishes market conditions that enable firms to leverage the economy's competitive advantage. This shows that the government is a critical role player in enhancing destination competitiveness (Hall, 2007; Kubickova & Martin, 2020;

Kubickova, 2017). Consequently, destination competitiveness must be viewed just like other public goods (Kubickova & Martin, 2020). This is because the government plays a key role in assisting tourism growth to boost the economy and increase the quality of life of destination residents (Kubickova & Martin, 2020; Kubickova, 2017). Therefore, destinations with governments that can provide quality public goods have always been the most chosen ones among tourists (Croes, 2011; Woyo, 2018).

The government's involvement in tourism development is frequently reflected in public policies and the creation of government departments, ministries and institutions to manage the destination (Hall & Campos, 2014; Mei et al., 2015). However, poorly conceptualised policies introduced by governments in a destination have a negative effect on tourism competitiveness (Manuela & de Vera, 2015). This is one reason why the nexus between government initiatives and competitiveness attracts a lot of attention from academics (Kubickova et al., 2017) and practitioners.

To achieve the desired competitiveness in a destination, stakeholder cooperation is critical, and governments must play a much bigger role by creating an enabling environment (Dwyer et al., 2016; Hall, 2007; Woyo, 2018; Woyo & Slabbert, 2021). Furthermore, creating a competitive destination requires a balanced approach to stakeholder cooperation (Devine & Devine, 2011) to avoid the illusion that normally happens when tourism development is left entirely to the private sector (Karnani, 2011). Leaving the management of a destination to "*the private sector could result in the unbalanced development of infrastructure and market expansion, with the risk of growing congestion and increased pressure on environmental resources*" (Devine & Devine, 2011, p.1253). Therefore, government involvement is crucial in helping prevent activities that affect the welfare of residents.

Government involvement in tourism is not without criticism. On the one hand, there is a school of thought that argues that government policies are vital to the industry's success (Glaeser & Shleifer, 2002; Mei et al., 2015; Yang et al., 2006). On the other hand, some argue that public-sector involvement in tourism is excessive. Government involvement is bad for business because of its bureaucratic tendencies and blatant lack of transparency (Bull, 1995; Croes, 2011; Dolfsma, 2013; Mei et al., 2015); furthermore, it causes market distortion (Croes, 2011). However, the government remains a critical stakeholder in tourism development (Dwyer et al., 2016; Javed & Tučková, 2020; Kubickova & Martin, 2020) mainly due to *"its impact on any national economy and residents' quality of life"* (Javed & Tučková, 2020:997). Consequently, measuring these aspects in a country with ongoing political and economic crises (Woyo & Slabbert, 2020) is required.

The role of government in tourism development and competitiveness

The role of government in the tourism industry has long been recognised in the literature (Dwyer et al., 2016), as presented in Table 9.1. As stated earlier, its role is frequently reflected by developing national tourism policies, destination marketing campaigns, and special agencies for tourism (Kubickova & Martin, 2020). It is widely established that government plays an important role in tourism competitiveness (Devine & Devine, 2011; Lee, 2015). Based on this, developing a competitive tourist destination directly results from having a competent bureaucracy (Lee, 2015). However, it is important to note that the function is not universal and varies from country to country (Kubickova & Martin, 2020). Furthermore, tourism development is often in stages and thus requires different levels of government involvement. Destinations generally start with low government involvement and move towards higher participation to create an environment for tourism to compete globally. The level of involvement tends to decline once tourism has achieved a strong global industry presence. Consequently, it is imperative for this current study to establish the suppliers' perceptions and establish government involvement in Zimbabwe and how it influences competitiveness.

Methods

Zimbabwe is a tourist destination operating under distress for more than two decades (Woyo & Slabbert, 2020, 2021). The destination developed national tourism strategies through the Ministry of Tourism and Hospitality Industry, launched destination marketing campaigns and a tourism policy. These government interventions were motivated by the decline in tourist arrivals which emanated from the contested land reform programme and disputed elections (Woyo & Slabbert, 2020, 2021; Woyo, 2018). However, research understanding these government initiatives' role in enhancing a destination's competitiveness in distress is yet to emerge on a local level. On a global level, limited studies investigate the political environment and tourism competitiveness (Pike & Page, 2014), especially in countries with ongoing political and economic challenges like Zimbabwe.

There is a lack of research regarding government initiatives' role in tourist destinations' competitiveness (Kubickova, 2019), especially in Africa's distressed contexts. Exploratory research was deemed appropriate for the study and identifying variables that could be useful for quantitative research. A qualitative approach was selected because it helps investigate all aspects and how the supply side perceives the role that government initiatives have played on destination competitiveness in the last two decades. The government's role in destination competitiveness in a country with ongoing political and economic challenges presents new challenges that need to be identified to obtain critical insights (Strauss & Corbin, 1988).

An inductive qualitative approach was adopted using open-ended questions for in-depth online interviews (in English). The online interviews directly resulted from the need to follow existing health and safety regulations that were gazetted to contain the COVID-19 virus. The study employed non-probability sampling, and participants were invited to participate in the study via emails between October and November 2020. The final sample had a total of 15 participants that were drawn from tourism establishments operating in Zimbabwe (Table 9.2). Samples ranging between 14 and 30 are adequate for scientific analysis in tourism studies (Phelan, 2015; Yap and Ineson, 2009). Therefore, the sample of 15 was deemed consistent with previous influential qualitative research recommendations. It is advised that sampling should continue until the study reaches saturation, and saturation was reached after the 10th interview. Fifteen interviews ranging from 36 minutes to 40 minutes in length were conducted, which was deemed sufficient for this exploratory study.

The key criteria for inclusion in the study were for the participant to hold a formal management position in the tourism and hospitality industry and have worked in the sector for at least ten years. Data collected in this study include demographic aspects of the participants, their perceptions of government initiatives, and how they have influenced Zimbabwe's competitiveness as a destination post-2000. All participants were asked four main questions and additional probing questions to solicit more answers. Permission to record participants was sought before the interviews and was granted. All interviews were recorded on the Zoom cloud facility and were transcribed verbatim before being used for the study.

Content analysis was used to analyse the data, and this resulted in several sub-themes from the questions administered during the interviews. The interview responses were read and re-read several times to interpret variations and discern similarities and differences in the content. Similarities and variations observed resulted in numerous underlying sub-themes that helped reorganise data and identify reoccurring themes (Thyme et al., 2013; Mazaheri et al., 2013). Responses were coded to uncover similarities (Table 9.2). Careful observation of the sub-themes resulted in five dominant themes labelled funding, policy issues, tourism framework, enabling environment and budget incentives (Figure 9.1).

Table 9.2: Participants' profile

Participant	Position/Department	Type of organisation	Age	Experience (in total)	Location
S1	Operations Manager	5 Star Hotel	48	15	Harare
S2	Marketing Manager	Tour Operator	38	11	Victoria Falls
S3	Finance Officer	Car Rental	45	18	Kariba
S4	Finance and Administration	National Park	57	25	Hwange
S5	General Manager	National Tourism Organisation	49	12	Harare
S6	General Manager	4 Star Hotel	50	20	Victoria Falls
S7	Executive Housekeeper	3 Star Hotel	46	15	Harare
S8	Services Manager	Low-cost Airline	38	10	Harare
S9	Food and Beverage Manager	5 Star Hotel	40	16	Masvingo
S10	Sales Manager	Tour Operator	40	15	Harare
S11	General Manager	Tour Operator	39	10	Mutare
S12	Manager	Restaurant	35	10	Nyanga
S13	Manager	Tour Operator	37	15	Victoria Falls
S14	Food and Beverage	Guest lodge	43	20	Victoria Falls
S15	Finance and Administration	Arts organisation	48	24	Harare

9

Findings and discussion

The responses provided a rich overview of the tourism suppliers' perceptions in a destination with ongoing political and economic challenges. This chapter identifies and analyses the five key themes that underpin government initiatives' role and their perceived influence on destination competitiveness in a distressed destination. Funding, policy issues, tourism framework, enabling environment, and budget incentives, each with several distinctive dimensions expressed as their sub-themes, are presented in the map of themes and sub-themes in Figure 9.1.

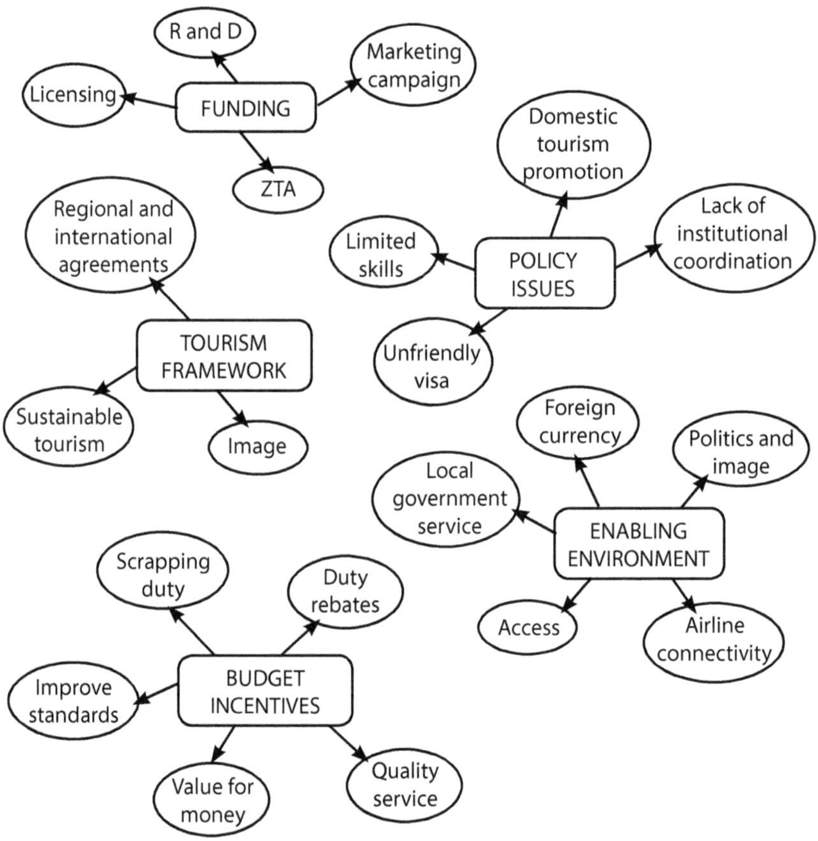

Figure 9.1: Supplier's perceptions of government role in tourism competitiveness

The most prominent theme that emerged was related to the government's incentives through the National Budgets since 2000. Though argued to be erratic, most of the participants recognised that government has since 2000 played an important role in enhancing destination competitiveness in Zimbabwe. The incentives through rebates on importing capital goods for tourism projects, scrapping of duty on safari and car hire vehicles gave Zimbabwe a critical boost during a time of tourism decline, especially after the year 2008.

Some critical reflections from industry experts are worth noting. On the question of the importance of incentives that were announced through the national budgets in the last 20 years, participant S12 voiced out that:

> *"The extension of the tourism rebate regulation applies to the import of capital goods for tourism projects and is a significant incentive for those involved in tourism to take the opportunity to upgrade and refurbish their products, properties and associated projects".*
>
> *"The scraping of duty on importing safari and car hire vehicles will boost the industry's productivity. Quality of equipment will give the clients value for money, impeccable service and this, in turn, will improve industry standards."*

Participants shared mixed opinions about the incentives by scrapping their duty and rebating on capital goods for tourism projects. Some lauded the Zimbabwean government's role in enhancing tourism competitiveness by enabling suppliers to acquire capital goods. However, others expected the government to provide a more enabling environment that makes it easy for the tourism industry to acquire foreign currency to buy capital goods. Though some participants indicated that rebates and scrapping of duty are significant on their profits and the determination of a fair price to the customer, recent studies continue to reflect that Zimbabwe is an expensive tourist destination (Woyo, 2018. Woyo & Woyo, 2019; Woyo & Slabbert, 2019, 2020; Woyo, 2021). Zimbabwe's pricing regime is too serious to be overlooked. It affects the price competitiveness and tourism recovery for both the domestic market (Woyo, 2021) and the international market (Woyo, 2018; Woyo & Slabbert, 2020). Therefore, while the government has helped provide incentives, more is required to influence the destination's price competitiveness.

Though conclusions regarding the impact of government spending have been mixed in past studies, most of the studies argue that government expenditure impacts economic development (Berggren, 2003) and destination competitiveness (Kubickova, 2019). Funding by the central government remained a recurring response under various categories of inquiries. While many of the participants agreed that Zimbabwe has world-class attractions, the industry's contribution to GDP remains relatively small because the country is still struggling with basic issues such as infrastructure, especially transport. This is generally exacerbated by the fact that Zimbabwe is still perceived as politically unstable (Woyo & Slabbert, 2020, 2021). Due to this, Zimbabwe is not able to secure funding from international markets for infrastructural development. Due to the political climate and economic environment prevailing in Zimbabwe at the moment, international sourcing partners for hospitality and tourism investment remains difficult to find.

Investment in tourism infrastructure is important for enhancing destination competitiveness (Woyo, 2018; Woyo & Slabbert, 2021; Xie & Tveterås 2020). Based on the analysis of the results, it was noted that the government in distressed contexts must continuously provide more funding for investing in tourism infrastructure (Xie & Tveterås 2020) to enhance tourist destinations' competitiveness. However, this is not currently happening in Zimbabwe. Furthermore, other issues raised in the study include increased funding for the national tourism organisation to conduct its duties, such as licensing and marketing campaigns. Due to economic challenges, it appears these aspects are not perhaps the priority of the government at the moment.

The third theme focused on enabling the environment for enhancing competitiveness in distressed contexts. Key issues raised included the lack of local government services, foreign currency, politics and image, price competitiveness and limited airline connectivity. Many participants argued that the government's role in creating an enabling business environment where tourism could thrive was limited. Past studies (Woyo, 2018; Woyo & Slabbert, 2020, 2021) note that Zimbabwe's competitiveness is affected by the general lack of direct flights from source markets. Poor air transport and road infrastructure is affecting Zimbabwe's competitiveness and there is a need for government to invest in air and road transport. Investment in infrastructure is critical for attracting a larger growth of international travellers to the destination. Furthermore, an improved infrastructure network is imperative for the destination in that it creates an enabling business environment for tourism establishments. According to the World Economic Forum, such an environment is critical for improving Zimbabwe's competitiveness rankings that are currently low. Furthermore, most of the participants agreed that the government must also improve laws to attract foreign investment and avoid policy inconsistency, especially with its empowerment laws. Improving the laws could help attract investors in tourism infrastructure through public-private partnerships in the tourism industry.

Safety and security are classified in the literature as deterrents for travel (Farmaki et al., 2019). While Zimbabwe continues to receive arrivals, the government's role in ensuring safety and security has been questionable, especially after firing live ammunition to protesting citizens (Woyo & Slabbert, 2020). Managers in the sector are strongly advised to develop strategies that help ensure travellers' security and safety in countries with political instability. This is critical in improving the destination's brand image, which has been negative for a long time. Furthermore, aspects listed under policy issues such as visas, lack of coordination, and domestic tourism promotion require concerted stakeholder participation to enhance a destination's competitiveness.

Regarding the unfriendly visa regime, Zimbabwe is generally ranked 27th in the Africa Visa Openness report, and this could be affecting its overall tourism competitiveness ahead of competitors. Supplier participants argued that governments in distressed contexts need to facilitate tourists' free movement, which will become more critical post-COVID-19. Furthermore, the ease of doing business has been noted with great concern and it does not come cheap in Zimbabwe due to rampant corruption. Once tourists arrive in Zimbabwe, they may face other hurdles including several police roadblocks on the highways, which often discourages tourists from staying longer, affecting overall competitiveness.

Another theme identified was the tourism framework, and most of the participants highlighted that tourism is an important sector that cannot be operated without a guiding policy framework. The development of the tourism industry in Zimbabwe has been for a long time premised on the concept of low volume-high spending; tourism has been a narrative that had no policy guidelines, and the presence of policies is generally deemed crucial for competitiveness. However, the Ministry of Tourism and Hospitality Industry has completed the National Tourism Policy drafting to guide the country's tourism sector operations. Though the National Tourism Policy is now in place, most respondents argued it could have driven tourism development for improved competitiveness.

The vacuum created due to the absence of a policy affected creating an enabling tourism environment that is important for tourism competitiveness, especially in distressed contexts. Though the National Tourism Policy was designed to create an enabling environment for sustainable tourism development and community participation, this objective's realisation is difficult considering the many challenges that Zimbabwe is facing (Woyo, 2021).

New challenges and applications

The study shows that while Zimbabwean tourism suppliers generally agree that government plays a key role in enhancing competitiveness, several challenges emerge. First, government involvement appears to be fuelling more corruption, especially with initiatives such as scrapping the duty to import tourism equipment. This, therefore, tends to affect the influence that the government should have on destination competitiveness, especially when it comes to creating an enabling environment. Second, aspects such as heavy police roadblocks that characterise Zimbabwean highways (Woyo & Woyo, 2019) create negative destination images that tend to reduce the competitiveness of the destination and discourage long durations of stay. These roadblocks will also pose a challenge for restarting tourism following the global COVID-19 pandemic, where people are more comfortable travelling using private road transport. Third, there is a general lack of realisation from the government concerning what a competitive tourist destination will do for Zimbabwe. This is reflected in the long gestation period of national tourism policies and guidelines.

To summarise, policymakers must be mindful of the policies and initiatives that they implement. Furthermore, they must also be conscious of the effect these policies have on destination competitiveness. Too little and too much involvement can either amplify or dampen destination growth, thus, lowering competitiveness. Individual policies must be fully understood, monitored, and revisited for deregulation to be effective.

Conclusions

In conclusion, it is critical for governments, especially in countries with distress, to be aware of the policies and initiatives they put in place and their impact on the tourist destination's competitiveness. Contrary to

Javed and Tučková (2020) findings, this study concludes that a higher level of government involvement is required to enhance tourism competitiveness in distressed contexts. Therefore, a comprehensive understanding, monitoring, and evaluation of government initiatives using a stakeholder approach is imperative to ensure a destination's successful tourism competitiveness. Furthermore, this study shows that some initiatives like incentives through the scrapping of duty and duty rebates help sustain the tourism industry's competitive advantage when the economy is weak. Consequently, governments of destinations with ongoing crises, like Zimbabwe, must evaluate and decide which policies directly affect tourism and the economy.

The reader should use caution when interpreting the results of this chapter. The findings are specific to Zimbabwe as a country in distress and can only be generalised to another country with similar ongoing political and economic challenges. Furthermore, other research limitations could be noted. First, the sample was qualitative and did not cover the whole spectrum of the establishments making Zimbabwe's tourism sector. The study did not consider other tourism stakeholders' viewpoints, and this leaves a significant population gap for future researchers to consider other players in the tourism industry. Future studies should look at other sectors that were not represented in this study. Second, this study is based on qualitative analysis of data collected from managers of establishments, which may also not justify the generalisation of the findings. This chapter makes a point in giving an exploratory stance to trigger further empirical investigations of the government's role in enhancing tourism competitiveness of destinations in distress. Future researchers can take various variables exhibited as sub-themes in Figure 9.1 and operationalise them in investigations of the nexus between government initiatives and tourism competitiveness. Third, quantitative studies using primary data can also bring much richer and newer insights into government initiatives' impact on tourism competitiveness in distressed contexts.

Last, future studies should include different regions and countries in Africa and other continents, leading to different results. Addressing these limitations of the study could improve our understanding of the nexus between government initiatives and tourism competitiveness and help practitioners develop effective strategies to sustain competitiveness in a tourist destination.

Highlights

- ☐ The study measured the perceptions of suppliers operating in a distressed destination regarding the nexus between government initiatives and tourism competitiveness.
- ☐ It analyses government initiatives' role in tourism competitiveness using data from suppliers operating in a distressed destination.
- ☐ The government's involvement in the process of tourism development, especially in developing countries, is critical.
- ☐ Developing economies with ongoing political and economic challenges like Zimbabwe have not been comprehensively researched, especially on competitiveness and government roles.
- ☐ The role of the government is important for enhancing tourism competitiveness. A higher level of government involvement is needed for Zimbabwe to enhance competitiveness.

Questions

1. What is the nexus between the role of government and tourism competitiveness? How are government initiatives and competitiveness linked together?
2. What is the appropriate level of government involvement that can provide distressed destinations with high destination competitiveness levels?
3. Should the government intervene in tourism development? Which public services should the government provide for the tourism industry?

References

Abreu-Novais, M., Ruhanen, L. & Arcodia, C. (2016) Destination competitiveness: What we know, what we know but shouldn't and what we don't know but should, *Current Issues in Tourism*, 19(6), 492-512.

Armenski, T., Dwyer, L., & Pavluković, V. (2018). Destination competitiveness: Public and private sector tourism management in Serbia. *Journal of Travel Research*, 57(3), 384-398.

Berggren, N. (2003) The benefits of economic freedom: A survey. *The Independent Review*, 8(2), 193–211.

Bull, A. (1995) *Economics of Travel and Tourism,* 2nd edn, Longman Australia

Croes, R. (2011) Measuring and explaining competitiveness in the context of small island destinations, *Journal of Travel Research,* 50(4), 431–442.

Croes, R., & Kubickova, M. (2013) From potential to ability to compete: Towards a performance-based tourism competitiveness index, *Journal of Destination Marketing & Management,* 2(3), 146–154.

Cronjé, D.F. & du Plessis, E. (2020) A review on tourism destination competitiveness, *Journal of Hospitality and Tourism Management,* 45, 256-265.

Crouch, G. (2011) 'Destination competitiveness: An analysis of determinant attributes', *Journal of Travel Research,* 50(1), 27-45.

Devine, A. & Devine, F. (2011) Planning and developing tourism within a public sector quagmire: Lessons from and for small countries, *Tourism Management,* 32(6), 1253-1261.

Dodds, R. & Holmes, M.R. (2020) Is blue flag certification a means of destination competitiveness? A Canadian context, *Ocean & Coastal Management,* 192, 105192.

Dolfsma, W. (2013) *Government Failure: Society, markets and rules*. Edward Elgar, Cheltenham.

Dwyer, L., Dragicevic, V., Armenski, T., Mihalic, T. & Cvelbar, L.K. (2016) Achieving destination competitiveness: and importance-performance analysis of Serbia, *Current Issues in Tourism,* 19(13), 1309-1336.

Farmaki, A., Khalilzadeh, J. & Altinay, L. (2019) Travel motivation and demotivation within politically unstable nations, *Tourism Management Perspectives*, 29, 118-130.

Glaeser, E. L., & Shleifer, A. (2002) Legal origins, *The Quarterly Journal of Economics*, 117(4), 1193–1229.

Gooroochurn, N. & Sugiyarto, G. (2005) Competitiveness indicators in the travel and tourism industry, *Tourism Economics,* 11(1), 25-43.

Hall, C.M. (2007) Tourism and regional competitiveness', in J. Tribe, & D. Airey (eds.) *Advances in Tourism Research*, Elsevier, Oxford, pp. 217-230.

Hall, C.M. & Campos, M.J.Z. (2014) Public administration and tourism–international and Nordic perspectives, *Scandinavian Journal of Public Administration,* 18(1), 3-17.

Javed, M. & Tučková, Z. (2020) The role of government in tourism competitiveness and tourism area life cycle model, *Asia Pacific Journal of Tourism Research,* 25(9) 997-1011.

Karnani, A. (2011) Doing well by doing good. The Grand Illusion, *California Management Review,* 53(2), 69-86.

Kubickova, M. (2016) The role of government in tourism: Linking competitiveness, freedom, and developing economies, *Czech Journal of Tourism,* 5(2), 73–92.

Kubickova, M. (2019) The impact of government policies on destination competitiveness in developing economies, *Current Issues in Tourism*, 22(6) 619-642.

Kubickova, M., Croes, R., & Rivera, M. (2017). Human agency shaping tourism competitiveness and quality of life in developing economies. *Tourism Management Perspectives*, 22, 120-131.

Kubickova, M., & Li, H. (2017). Tourism competitiveness, government and tourism area life cycle (TALC) model: the evaluation of Costa Rica, Guatemala and Honduras. *International Journal of Tourism Research*, 19(2), 223-234.

Kubickova, M. & Martin, D. (2020) Exploring the relationship between government and destination competitiveness: the TALC model perspective, *Tourism Management*, 78, 104040.

Lee, S. (2015) Quality of government and tourism destination competitiveness, *Tourism Economics*, 21(4), 881–888.

Leung, X.Y. & Baloglu, S. (2013) Tourism competitiveness of Asia Pacific destinations, *Tourism Analysis*, 18(4), 371-384.

Manuela Jr., W.S. & de Vera, M.J. (2015) The impact of government failure on tourism in the Philippines, *Transport Policy*, 43, 11-22.

Mazaheri, M., Eriksson, L.E., Heikkilä, K., Nasrabadi, A.N., Ekman, S.L., & Sunvisson, H., (2013) Experiences of living with dementia: qualitative content analysis of semi-structured interviews, *Journal of Clinical Nursing*, 22(21-22), 3032–3041.

Mei, X.Y., Arcodia, C. & Ruhanen, L. (2015) The national government as the facilitator of tourism innovation: Evidence from Norway, *Current Issues in Tourism*, 18(12), 1172-1191.

Phelan, K. V. (2015)'Elephants, orphans and HIV/AIDS, *Worldwide Hospitality and Tourism Themes*, 7(2), 127–140.

Pike, S. & Page, S.J. (2014) Destination Marketing Organizations and destination marketing: A narrative analysis of the literature. *Tourism Management*, 41, 202-227.

Ritchie, J.R.B. & Crouch, G.I. (1999) Tourism, competitiveness, and societal prosperity, *Journal of Business Research*, 44(3), 137-152.

Strauss, A. & Corbin, J. (1988) *Basics of Qualitative Research Techniques*. Sage, Thousand Oaks, CA.

Thyme, K.E., Wiberg, B., Lundman, B., & Graneheim, U.H., (2013) Qualitative content analysis in art psychotherapy research: concepts, procedures, and measures to reveal the latent meaning in pictures and the words attached to the pictures, *Arts Psychotherapy*, 40(1), 101–107.

Tučková, Z., & Jurigová, Z. (2014) *The Meaning of Tourism and Tourism Services in The V4 Countries*, s. 255. GEORG.

UNWTO. (2019) *International Tourism Highlights*. https://www.unwto.org/publication/international-tourism-highlights-2019-edition

Wan, Y.K.P. & Bramwell, B. (2015) Political economy and the emergence of a hybrid mode of governance of tourism planning, *Tourism Management*, 50, 316-327.

Webster, C. & Ivanov, S. (2014) Transforming competitiveness into economic benefits: Does tourism stimulate economic growth in more competitive destinations?' *Tourism Management, 40, 137-140.*

World Travel and Tourism Council. (2015) *Policy Impact Analysis*. Retrieved from http://www.tourismeconomics.com/economic-impact/policy-impact-analysis. Accessed 12 October 2020.

Woyo, E.(2018). An assessment of brand Zimbabwe's competitiveness and attractiveness as a tourism destination. PhD thesis, North West University, Potchefstroom.

Woyo E. (2021) The sustainability of using domestic tourism as a post-COVID-19 recovery strategy in a distressed destination, in W. Wörndl, C. Koo & J.L. Stienmetz (eds) *Information and Communication Technologies in Tourism* 2021. Springer, Cham.

Woyo, E. & Slabbert, E. (2019) Cross-border destination marketing of attractions between borders: The case of Victoria Falls, *Journal of Hospitality and Tourism Insight,s* 2(2), 145–165.

Woyo, E. & Slabbert, E. (2020) Unpacking the motivations, satisfaction and loyalty of tourists travelling to a distressed destination, *Anatolia*, 31(4),536-548.

Woyo, E. & Slabbert, E. (2021) Tourism destination competitiveness: A view from suppliers operating in a country with political challenges, *South African Journal of Economic and Management Sciences*, 24(1), 1-10.

Woyo, E. & Woyo, E. (2019) Towards the development of cultural tourism as an alternative for tourism growth in Northern Zimbabwe, *Journal of Cultural Heritage Management and Sustainable Development*, 9(1), 74–92.

Xie, J. & Tveterås, S. (2020) Economic decline and the birth of a tourist nation, *Scandinavian Journal of Hospitality and Tourism*, 20(1), 49–67.

Yang, L., Wall, G., & Smith, S. L. (2008) Ethnic tourism development: Chinese government perspectives, *Annals of Tourism Research*, 35(3), 751–771.

Yap, M.H. & Ineson, E.M. (2009) HIV-infected employees in the Asian hospitality industry, *Journal of Services Management*, 20(5), 503–520.

10 Extended Reality technologies as a tool for managing crises and shaping tourism safety perceptions

Christina Karadimitriou and Elina Michopoulou

Abstract

New technologies are considered by different industries as a useful tool for having efficient emergency and crisis management. For tourism industry in particular (that involves and interfaces with multiple other industries), it is critically important to act proactively in a risk situation, to effectively face a disaster, and to reduce the impact of a crisis. This chapter provides an overview of the Extended Reality (XR) technologies (Augmented Reality [AR]; Virtual Reality [VR]; Mixed Reality [MR]). It discusses opportunities of using XR in tourism, and it provides contemporary examples of XR applications. It also focuses on emergency management via XR in tourism. Finally, it provides specific recommendations for XR use before, during, and after a crisis in order to better prepare for, manage and recover from emergencies and crisis.

Keywords:

Tourism, crisis management, virtual reality, augmented reality, mixed reality, extended reality, tourism safety

Introduction

There is no definition of crisis which is globally accepted (Coombs, 2012). According to Pauchant and Mitroff (1992) the meaning of a crisis focuses on the disruption that has an impact on a specific system and constitutes threat for basic assumptions. Concerning tourism, Sönmez et al. (1994) noted that a crisis may have an impact not only on related businesses, but is also capable of destroying the destination. Thiessen (2011) suggests two typologies of crisis based on the dimensions: time and content. Considering the time dimension, Coombs (2012) proposes that a crisis includes three phases: pre-crisis, crisis and post-crisis. With regards to the content dimension, Rosenthal and Kouzmin (1993) consider intentional man-made crises (terrorist attack) and natural external causes (hurricanes), whilst James and Wooten (2005) focus on sudden (earthquake) or predictable crisis (recession). The notion of stability and safety in economic, health and social terms being a static condition, has been contested in the recent years (Pappas & Glyptou, 2021). Examples such as the 9/11 terrorist strikes in 2001, the Indian ocean tsunami catastrophe in 2004, the refugee crisis in 2015, and many others only prove that a serious crisis isn't something rare, but rather increasingly commonplace.

Considering the most recent example of the Covid-19 pandemic, it caused the slowdown of most economic activities, as well as the introduction of radical changes in global health and safety standards. As a result, many industries were affected, including transportation, educational institutions, and sport (Chakraborty & Maity, 2021); but tourism was particularly disrupted. While the impact of Covid-19 on tourism is yet to be fully understood, some facts remain unchallenged. According to the World Travel and Tourism Council lists, at immediate risk remain up to 75 million workers with the possibility the daily loss of up to one million jobs in the travel tourism sector. The potential Travel Tourism GDP loss was estimated in 2020 of up to US$ 2.1 trillion (Skare et al., 2021). These are just some of its effects with significant implications for customers, businesses and destinations across value and supply chains.

The lack of risk management for a pandemic situation forced individuals to rapidly formulate and apply disaster management plans, many of which included technological solutions. Compared to traditional communications methods, new technologies such as Virtual Reality (VR),

Augmented Reality (AR) and Mixed Reality (MR) not only provide a safe environment but also present individuals with the opportunity for interaction and engagement. Technologies such as VR, AR and MR offer a great opportunity for crisis management in pre-, during and post-crisis phases, as they can assist in restoring feelings of safety and security but also revamping tourism and hospitality companies' and destinations' images.

While the existing literature for XR technologies applications in many fields such as education (Merchant et al., 2014; Cihak et al., 2016; Loureiro & Bettencourt, 2014), health care (Freeman et al., 2017; Cobbett & Snelgrove-Clarke, 2016; Alvarez et al., 2011, and retail (Park & Yoo, 2020; Fan et al., 2020; Bonetti et al., 2018) is abundant, the literature concerning the use of XR technologies in tourism, and especially as a tool for managing crises and shaping tourism safety, is predominantly silent. It is therefore necessary to further research the ways that all these new technologies can be valuable in avoiding or countering tourism crises. Hence, this chapter examines the use of AR, VR and MR technologies in the pre-, main, and the post-crisis phases and discusses potential applications for the tourism and hospitality industries. The theoretical contribution of the chapter lies within the fact that it jointly considers three distinct knowledge areas and explores issues that exist in the intersection of technology, crisis management and tourism. To do so, the chapter first discusses the different XR technologies, in particular AR, VR and MR. It then proceeds to contextualize these technologies within the tourism domain, providing some current application examples. It proceeds to review the use of XR technologies as emergency and crisis management tools. The chapter then continues by considering XR technologies for managing tourism crises. Finally, it offers a summary of key point and concluding remarks.

XR technologies

According to Fast-Berglund et al. (2018) the term extended reality (XR) refers to all combined environments (real and virtual) and includes different types of technologies i.e Augmented Reality (AR), Virtual Reality (VR) and Mixed Reality (MR).

AR technology offers the user the opportunity to overlap virtual objects to the real world (Peddie, 2017). This is done through a variety of devices, usually grouped into three categories: (i) the Head Mounted Displays (HMD) which includes the optical see-through form and the video see-through (Karadimitriou & Trichas, 2020), (ii) systems where the real world is joined with through a coupled camera and a connected computer projecting the virtual image in the screens such as smartphones and tablets (Nishihara, 2015), and (iii) the Spatial Augmented Reality (SAR) which allows the user to interact with the virtual object by utilizing projectors which directly project digital information onto the objects of the real world (Azuma et al., 2001). According to He et al. (2018), AR technology is considered as a revolutionary invention and its use has increased rapidly over the past few years. AR can be very effective in enabling interaction with the real environment (Tussyadiah et al., 2018b). The most popular application which brought AR technology to the general public was Pokémon Go (Rauschnabel et al., 2017). The AR market is expected to increase from US$5.91 billion to US$198 billion by the year 2025 (Shah, 2019).

VR technology can be defined as an immersive-interactive technology in which the users immerse into a digital environment and interact with the digital elements. (Milgram & Kishino, 1994; Zhou, 2021; Zhang, 2021). Visual immersion according to Goedert and Rokooei (2016) and Bowman and McMahan (2007) is defined by how similar a system's visual output is to reality. VR technology has many applications in different fields such as military training (Chen et al., 1999) entertainment (Zyda, 2005), education (Merchant et al., 2017), research (Meißner et al. 2017), retail (Bonetti et al., 2018) and healthcare (Freeman et al., 2017), due to the immersiveness that it offers (Berg & Vance, 2016). Lucas et al. (2007) note that VR technology is a very useful and safe alternative when individuals must be trained in dangerous occasions. It is also very effective in phobia therapy (Cohen et al., 2005) and in speech therapy (Goedert & Rokooei, 2016). According to Dascal et al. (2017), VR technology in hospital inpatients was not only safe but also increased customer satisfaction factor in different medical settings. VR content can be displayed through desktop, Cave Automatic Virtual Environment (CAVE) and HMD (Manjrekar et al., 2014). It is worth mentioning that in 2019 the VR market size was valued at US$10.32 billion and the expectation of growing rate is 21.6% from 2020 to 2027 (Market Analysis Report, 2021).

In MR technology (unlike VR) virtual objects and the real world are merged (Pan et al., 2006; Tamura et al., 2001; Syal & Mathew, 2020).

The basic characteristic of MR is visual coherence (Collins et al., 2017). Essentially in MR the user can interact with both real and digital elements, with the differentiating factor being that the elements can also interact with each other. Physical factors, imaginary factors and virtual factors are contributing to the final MR experience involving multisensorial responses such as hearing, sight etc. (Konttinen, 2005). The main difference between the XR technologies is that in AR the user has direct contact with the physical environment and remains present. In VR technology the user experiences a full digital environment and is totally cut off from the surrounding environment, whereas with MR technology the user coexists with the digital and the real word.

XR technologies in tourism

AR technology has had a great impact on the tourism industry particularly as a valuable marketing tool (Ukpabi & Karjaluoto, 2016; Sigala, 2018). AR technology offers tourists the opportunity to enhance their experience by exploring many unfamiliar sights (Han et al., 2014). This is particularly true within a cultural heritage tourism context (Cranmer et al. 2016) as, for instance, museums tend to use this technology in order to be more engaged with the visitor and at the same time provide a much more interactive content (Scarles et al., 2016). Tussyadiah et al. (2018a) argue that AR enables users to create memorable experiences. One of the most impressive examples of AR applications in hotel settings was launched by Holiday Inn for the London 2012 Olympic and Paralympic Games. Guests were offered the experience of seeing Britain's 2012 Olympic and Paralympic athletes in action, on screens in bedrooms, halls and in the reception (Innotour, 2020). Another example was introduced by the Hub Hotel from Premier Inn which used AR technology as a tourist information tool. Using a wall map in their hotel rooms, the tourists have the opportunity through a tablet or a smartphone to have access to extra information about local places (Revfine, 2020)

VR technology is becoming increasingly important for the tourism industry (Griffin et al., 2017). According to Kim et al. (2020), VR technology is shown to have a positive effect on tourist behaviour and having a VR experience is more likely to increase tourists' desire to participate in an activity (Jeng et al., 2017; tom Dieck et al., 2019). More importantly, VR technology has the potential to increase the desire of the tourist to visit a destination (Huang et al., 2016). According to Guttentag (2010), VR environments can be used not only as a promotional tool for destinations by immersing the consumers in new tourism experiences, but it can also be used to educate tourists to respect and protect the attractions, especially the sensitive ones. It is worth mentioning that VR and MR technologies are used by second chance tourism to assist destinations or sights that are deteriorated or destroyed by helping them to be "reborn" (Bec et al., 2021). A great example of VR technology in museums is the 'Mona Lisa: Beyond the Glass' VR project. In 2019 the Louvre (Paris) launched 'Mona Lisa: Beyond the Glass', an experience which was based on VR technology and explored the painting as a part of a blockbuster exhibition dedicated to Leonardo da Vinci. Employing interactive design, animated images and sound, the visitors of the museum had the opportunity to discover more details concerning Mona Lisa painting such as the texture of the wood panel and the effect of time in the appearance (Coates, 2020).

The dynamic field of XR technologies offers the tourism industry great marketing and promotion opportunities across the customer journey through various mechanisms such as gamification (Parapanos & Michopoulou, 2019), while evidence increasingly indicates that XR technologies will become more entangled to different tourism aspects. For instance, Flavian et al. (2019) examined the impact of VR on the customer purchase journey; Tussyadiah et al. (2018a) focused on the attitude and the behaviour change toward destinations; and Disztinger et al. (2017) and Huang et al. (2016) looked into the acceptance of VR technology for travel planning. Hence, Extended Reality enables companies and destinations to interact and engage with tourists particularly before and at the destination, enabling value co-creation for all stakeholders (Cavagnaro et al., 2021).

Emergency management via XR technologies

As mentioned previously, emergency management includes three main phases, pre-crisis, crisis and post-crisis (Coombs, 2012). The complexity of the phases of crisis management demands meticulous preparation and organisation in order to avoid escalation to a bigger crisis with much more serious consequences. XR technologies are useful in this way as they can assist with increasing preparedness in the pre-crisis phase, responsiveness during crisis and rejuvenation post-crisis.

VR, AR and MR technology can be used for emergency management, especially in the pre-crisis phase, offering several potential benefits. According to Lin et al. (2019), it is very difficult to create in real time a disaster scene because of legal and ethical restrictions. With XR technology it is much easier to create a disaster scene and especially due to the immersiveness of Virtual Reality it is easier to simulate a disaster and avoid any real risk for the participants.

At the same time AR, VR and MR can assist in developing emergency communication plans by identifying critical aspects to be considered by these plans that would otherwise not be apparent without prior experience. Hence, these technologies have become mainstream research methods (Lin et al., 2020) as they provide control over the variable parameters and disaster factors which influence emergency management (Lin et al., 2020). In terms of training, for instance, with XR technologies participants experience an interactive training which is easier to memorise and therefore develop better safety knowledge, enabling participants to have a similar response in a real-world crisis situation (Li et al., 2018). According to Meng and Zhang (2014), to improve the quality of an emergency simulation in a VR experiment, additional external environmental factors should be added. For example, a VR environment with fire and smoke is likely to induce higher physiological and psychological stress upon the participants, and therefore generate an improved stimulation of a real fire condition (Smith & Trenholme, 2009). XR technologies can also successfully simulate search and rescue work to train individuals. Lochhead and Hedley (2019) used not only AR but also MR technology to create simulations of evacuation in an actual environment of a building.

The post-crisis phase (also considered as a recovery phase) is a crucial phase due to the importance and urgency of rebuilding material structures and restoring feelings of safety of directly and indirectly involved individuals (Pappas, 2019). According to Wiederhold and Bouchard (2014) and Jang et al. (2000), VR technologies are helpful in addressing anxiety disorder due to the immersion factor. Designing a customizable environment according to the user needs and creating a scene that is close to the one that traumatized the person, provides an invaluable tool to the procedure of treating trauma. Pourmand et al. (2017) also suggest that new immersive technologies can be successfully applied to treatment of acute pain and be received as a supplementary psychological support. XR technologies may also be used as a visualization tool to process building information after emergencies such as for damage reconnaissance and reconstitution process (Matini et al., 2019).

XR technologies for managing tourism crises

Considering that destinations are amalgamations of tourism products and services (Buhalis, 2000), any potential crisis affects everyone within their respective value and supply chains, although admittedly to varying extents. Hence, it is important for all tourism stakeholders to prepare for the unavoidable fact that a crisis will occur sooner or later, the nature of which is unknown and unpredictable and therefore imbued with the promise of disruption. This very unpredictability of crises demands that emergency plans are in place to ensure minimal disruptions. XR technologies can assist tourism organisations and destinations to prepare for, respond to and recover from a crisis by using different applications appropriate for each phase of the crisis: before, during and after. As technologies develop and are adopted by the tourism sector, paving the way to entire e-tourism ecosystems and smart destinations (Buhalis, 2020), the use and adoption of XR technologies becomes easier due to interoperability and other synergistic effects with existing technological infrastructures. Berglund et al. (2020) suggest that smart environments with XR technology will allow a better response in emergency situations.

In terms of pre- and during crises phases, XR technologies and especially AR and VR can be used in several ways. Whilst the applications of these technologies primarily concern tourism suppliers and less so the tourists, tourists should in no way be excluded from the picture. Both technologies (especially VR), due to their immersive advantage, can simulate an emergency situation before it happens, so that tourism and hospitality staff (or tourists) can assess their ability to perform appropriately under the circumstances.

According to Perlman et al. (2014), using VR can be more helpful than traditional methods (such as photographs) for the participants to understand the emergency situation, as they allow participants to experience an emergency scene before real emergencies occur. Exposure to a simulated environment can introduce some sort of familiarization and therefore yield a more structured and organised reaction, upon which an effective crisis management plan can be designed. Such simulated digital and safe environments are more effective to instill emergency-handling knowledge to learners (Lovreglio et al., 2018). Even if the technology for both tourists and tourism suppliers is the same, the content will be different depending on their needs and the phase of the crises; for example, *pre-crisis* can focus on hazard recognition and prevention or safety training, *during crisis* can focus on human evacuation or search and rescue (Zhu & Li, 2021). During the pre-crisis phase, VR technology can focus on training simulation systems (i.e. evacuation, contact and supporting families and victims, front-line spokesperson for the crisis) within an immersive environment, which will include many different stages and layers of difficulty to which the user can react by taking ad hoc decisions and evaluate their impacts. Through this process, users can define the most suitable contingency plan whilst they can also formulate alternative ones. Concerning AR technologies, there can be a specific application enabling GPS with the use of tablets and smartphones in order to help tourists before a potential disaster, to define the safety of streets and buildings and to progress to a specific evaluation format during an emergency.

An example of XR technologies during crises is the use of drones with AR capabilities. These can be used to collect real time data, provide visuals with street names or building information overlays at a resort/destination to assess for instance the extent of damage, movement bottlenecks and support on the spot decision-making. VR technology in this phase can be also used as an effective supporting management tool for all the people that are working under high pressure environment conditions and for the families of the victims that will be under stressful conditions. Anxiety and emotional fatigue could be reduced through a software that will use positive memories and an immersive calmness environment.

Hence, XR technologies offer prime opportunities, for instance to hoteliers, event managers or attraction sites to better prepare for and respond to different types of crisis (i.e. fire, flood, earthquake or terrorist attacks) and test and employ different procedures (i.e. health and safety protocols, evacuation routes or communication chain of command).

The use of XR technologies for the post-crisis phase is even more critical for tourism, as the recovery phase includes not only the restoration of physical space (i.e. building reconstruction or damage detection) but also the repositioning of the destination image (or rebranding) in the minds of locals and tourists. Central to achieving the latter is the ability to restore feelings of safety, and trust that the destination's new condition resembles its pre-crisis condition. Unless this positive attitude is embraced and displayed by the locals and accepted by the tourists any attempts to stimulate tourism demand are futile. VR technology can also help to treat post-traumatic stress disorder (Geraldi et al., 2010).

With regards to the physical space, combinations of AR, VR and MR technologies can also be used to digitally 'rebuilt' a damaged or destroyed attraction site or heritage monument and recreate the environment. For example, Toubekis et al. (2009) with VR technology allowed tourists to see and interact with the destroyed Buddha figures in Afghanistan. The Roman Baths in Bath UK also developed an AR mobile app to allow visitors to experience reconstructions of the Baths at key moments in history, as part of their 5G Smart Tourism project (Thomas et al., 2019). Hence, XR technologies may be used not only to enhance the tourism experience (Beck, Rainoldi, & Egger, 2019; Guttentag, 2010; Wei, 2019) but also as a strategic emergency management tool.

Conclusions

Crises are inevitable and, as witnessed over the last few years, they tend to occur alarmingly more frequently. Dealing with crises is therefore an unavoidable reality that tourism businesses as well as destinations need to come to terms with. Fast and accurate decision-making is paramount when handling a crisis as it may be the very thing that ensures survival, in some cases literally. Hence, the examination of different scenarios and the development of emergency management plans is not optional. It is essential to maximise preparedness pre-crisis, optimize responsiveness during crisis, and minimise negative impacts including perceived safety of the destination or organisation post-crisis.

Extended reality technologies could be a great assistive tool in all the crisis management stages. For example, with XR technologies an emergency scene can be produced with safety and the mock-experience can be repeated indefinitely. This allows participants to spend as much time as is needed to examine all the possible complications that can occur and develop an appropriate plan to avoid or face them. This technology enabled mock experience can be offered to multiple participants, allowing a greater number of people to acquire the relevant skills and knowledge not just to prevent but to also face a crisis (i.e. the totality of a hotel's staff could be trained rather than few that would lead the others). The use of the right type of XR technology or a combination of them could be vital to control the disaster and to tackle issues that were not possible to foresee in the first stage. XR technologies can therefore be used as a strategic tool that cannot only minimise the negative effects of a crisis on a destination in the short term, but also enhance the destination image to project readiness and safety in the long run.

Yet, as new technological developments come to light, new applications will become available to facilitate crisis management. Further research is therefore required to address a number of issues including for instance: how to assure technology acceptance from tourism stakeholders; which technologies are more appropriate and useful mapped across the three stages of crisis; and in what ways does destination smartness enable crisis management.

In conclusion, it is important to examine XR technologies as quintessential tools in preparing for, handling and recovering from crises. This provides a new perspective on how to manage crisis within the tourism domain. By exploring the links between crisis management, tourism and XR technologies, it becomes clear that there are synergies that can assist tourism and hospitality to better cope with the challenging years ahead.

Highlights

- ☐ With the advance of XR technologies, new opportunities become available for emergency and crisis management in tourism and hospitality.
- ☐ Compared to traditional communication methods, XR technologies are more efficient in shaping safety feelings in tourism industry.
- ☐ In the near future, XR technologies are set to play a key role in enabling a more effective strategy for crisis and emergency management. XR technologies could be very helpful in accelerating recovery for tourism and hospitality after a crisis while they could play a vital role in pre-crisis phase due to the efficient training programs that can help identify and prepare for possible dangers.

Questions

1 What is the role of XR technologies in emergency and crisis management in tourism and hospitality industry?

2 What is the importance of XR technologies in shaping tourism safety perceptions?

3 In what ways XR technologies can contribute to gain the required knowledge in order to efficiently handle a crisis?

References

Alvarez, A.G., Dal Sasso, & G.T.M. (2011). Virtual learning object for the simulated evaluation of acute pain in nursing students. *Revista Latino-Americana de Enfermagem* 19, 229–237.

Azuma, R., Behringer, R., Feiner, S., Julier, S., & Macintyre, B. (2001). Recent advances in augmented reality. *IEEE Computer Graphics and Applications,* 21, 1–27.

Bec, A., Moyle, B., Schaffer, V. & Timms, K. (2021). Virtual reality and mixed reality for second chance tourism. *Tourism Management* , 83, (in press).

Beck, J., Rainoldi, M., & Egger, R. (2019). Virtual reality in tourism: A state-of-the-art review. *Tourism Review,* 74(3), 586–612.

Berg, L. P., & Vance, J. M. (2016). Industry use of virtual reality in product design and manufacturing: A survey. *Virtual Reality,* 21(1), 1–17.

Berglund, E.Z., Monroe J.G., Ahmed, I., Noghabaei, M., Do. J., Pesantez, J.E., Khaksar Fasaee, M.A., Bardaka,E.,Han,K., Proestos, G.T. & Levis, J.(2020). Smart Infrastructure, A vision for the role of the civil engineering profession in smart cities. *Journal of Infrastructure Systems*. 26.

Bonetti, F., Warnaby, G., & Quinn, L. (2018). Augmented reality and virtual reality in physical and online retailing. In: T. Jung, & M. tom Dieck (Eds.). *Augmented Reality and Virtual Reality. Progress in IS* (pp. 119–132). Cham: Springer.

10

Bowman, B.D., & McMahan, R.P. (2007). Virtual reality: how much immersion is enough? *IEEE Computer Society,* 40(7), 36–43.

Buhalis, D. (2000) Marketing the competitive destination of the future, *Tourism Management,* 21(1), 97-116.

Buhalis, D. (2020) Technology in tourism-from information communication technologies to eTourism and smart tourism towards ambient intelligence tourism, *Tourism Review,* 75(1), 267-272

Cavagnaro, E., Michopoulou, E., & Pappas, N. (2021) Revisiting value co-creation and co-destruction in tourism, *Tourism Planning & Development,* (in press)

Chakraborty, I. & Maity, P. (2021) COVID-19 outbreak: Migration, effects on society, global environment and prevention. *Science of the Total Environment.* 728, 138882

Chen, J.X., Dede, C.J., Fu, X. & Yang, Y., (1999). Distributed interactive learning environment, distributed interactive simulation and real-time applications.

Proceedings *3rd IEEE International Workshop on*, pp. 49–56.

Cihak, D. F., Wright, R. E., & Bell, S. M. (2016). Augmented reality as an instructional tool for teaching science vocabulary to postsecondary education students with intellectual disabilities and autism. *Journal of Research on Technology in Education*, 48(1), 38–56

Coates, C. (2020). *Virtual Reality is a big trend in museums, but what are the best examples of museums using VR?*. Museumnext. https://www.museumnext.com/article/how-museums-are-using-virtual-reality/. Accessed: 05/01/2021

Cobbett, S. & Snelgrove-Clarke, E. (2016). Virtual versus face-to-face clinical simulation in relation to student knowledge, anxiety, and self-confidence in maternal-newborn nursing. *Nurse Education Today*, 179–184.

Cohen, C.J., Hay, R., Urquhart, A., Gauger, P., & Andreatta (2005). A modular interactive virtual surgical training environment. *Interservice/Industry Training, Simulation, and Education Conference (I/ITSEC)*, p. 2074.

Collins, J., Regenbrecht, H., & Langlotz, T. (2017). Visual coherence in mixed reality *Presence Teleoperators and Virtual Environments*, 26(1), 16–41.

Coombs, T. (2012). Crisis communication and its allied fields. In: Coombs T., & Holladay S. (Eds.), *The Handbook of Crisis Communications* (pp 54-64). New Jersey: Blackwell Publishing.

Cranmer, E., Jung, T., & tom Dieck, M.C. & Miller, A. (2016). Understanding the acceptance of augmented reality at an organisational level: The case of Geevor Tin Mine Museum. In: A. Inversini, & R. Schegg (Eds.). *Information and Communication Technologies in Tourism* (pp. 637–650). Heidelberg: Springer.

Dascal, J., Reid, M., IsHak, W. W., Spiegel, B., Recacho, J., Rosen, B. & Danovitch, I. (2017). Virtual reality and medical inpatients: A systematic review of randomized, controlled trials. *Innovations in Clinical Neuroscience*. 14(1–2), 14–21.

Disztinger, P., Schlogl, S., & Groth, A. (2017). Technology acceptance of virtual reality for travel planning. *Information and Communication Technologies in Tourism*, 255–268.

Fan, X., Chai, Z., Deng, N. & Dong, X. (2020). Adoption of augmented reality in online retailing and consumers' product attitude: a cognitive perspective. *Journal of Retailing and Consumer Services*. 53, 101986-101986.

Fast-Berglund, A., Gonga. L & Lia, D. (2018) Testing and validating Extended Reality (xR) technologies in manufacturing. *Procedia manufacturing*. 25, 31-38

Flavian, C., Ibanez-Sanchez, S., & Orús, C. (2019). The impact of virtual, augmented and mixed reality technologies on the customer experience. *Journal of Business Research*, 100, 547–560.

Freeman, D., Reeve, S., Robinson, A., Ehlers, A., Clark, D., Spanlang, B., & Slater, M. (2017). Virtual reality in the assessment, understanding, and treatment of mental health disorders. *Psychological Medicine*, 47(14), 2393–2400.

Gerardi M, Cukor J, Difede J, Rizzo A & Rothbaum BO.(2010). Virtual reality exposure therapy for post-traumatic stress disorder and other anxiety disorders. *Current Psychiatry Reports*, 12(4):298–305.

Goedert, J.D. & Rokooei, S., (2016). Project-based construction education with simulations in a gaming environment. *International Journal Construction Education Research*. 12 (3), 208–223.

Griffin, T., Giberson, J., Lee, S. H., Guttentag, D., Kandaurova, M., Sergueeva, K., & Dimanche, F. (2017). Virtual reality and implications for destination marketing. *48th Annual Travel and Tourism Research Association, International Conference*, Quebec City, Canada.

Guttentag, D. A. (2010). Virtual reality: Applications and implications for tourism. *Tourism Management*, 31(5), 637–651.

He, Z., Wu, L., & Li, X. (2018). When are meets tech: The role of augmented reality in enhancing museum experiences and purchase intentions. *Tourism Management*, 68, 127–139.

Huang, Y. C., Backman, K. F., Backman, S. J., & Chang, L. L. (2016). Exploring the implications of virtual reality technology in tourism marketing: An integrated research framework. *International Journal of Tourism Research*, 18(2), 116–128.

Innotour (2020). *Augmented hotel experiences.* Available from: https://www.innotour.com/innovation-cases/augmented-hotel/.Accessed: 02/02/2021

James, E. H., & Wooten, L. P. (2005). Leadership as (un)usual: How to display competence in times of crisis. *Organizational Dynamics*, 34(2), 141-152.

Jang, D.P., Ku, J.H., Shin, M.B., Choi, Y.H., & Kim, S.I. (2000). Objective validation of the effectiveness of virtual reality psychotherapy. *CyberPsychology & Behavior*. 3(3), 369–374.

Jeng, M.-Y., Pai, F.-Y., & Yeh, T.-M. (2017). The virtual reality leisure activities experience on elderly people. *Applied Research in Quality of Life*, 12(1), 49–65.

Karadimitriou, C. & Trichas, N. (2020). Virtual and Augmented realities in emerging and advanced markets. In: Koutra C & Ahmad H.I (eds), *Emerging Human and Techno-Human Business Management Dynamics in a Globalized Environment* (pp.61-80) , NOVA

Kim, M.J., Lee, C.K., Jung, T., 2020. Exploring consumer behavior in virtual reality tourism using an extended stimulus-organism-response model. *Journal of Travel Research*. 59 (1), 69–89.

Konttinen, J., Pattanaik, S., & Hughes, C. E. (2005). *Real-Time Illumination and Shadowing by Virtual Lights in a Mixed Reality Setting*. School of Computer Science, Orlando:University of Central Florida

Li, X., Yi, W., Chi, H.L., Wang, X., & Chan, A.P.C. (2018) A critical review of virtual and augmented reality (VR/AR) applications in construction safety, *Automation in Construction*, 86, 150–162.

Lin, J., Zhu, R., Li, N., & Becerik-Gerber, B. (2020). How occupants respond to building emergencies: a systematic review of behavioral characteristics and behavioral theories, *Safety Science*, 122, (in press).

Lin, J.,Cao, L. & Li, N. (2019). Assessing the influence of repeated exposures and mental stress on human wayfinding performance in indoor environments using virtual reality technology. *Advanced Engineering Informatics*, 39, 53-61.

Lochhead, I. & Hedley, N. (2019). Mixed reality emergency management: bringing virtual evacuation simulations into real-world built environments. *International Journal of Digital Earth*. 12, 190–208

Loureiro, A., & Bettencourt, T. (2014). The use of virtual environments as an extended classroom–A case study with adult learners in tertiary education. *Procedia Technology*, 13, 97–106

Lovreglio, R., Gonzalez, V., Feng, Z., Amor, R., Spearpoint, M., Thomas, J., Trotter, M.,& Sacks, R. (2018). Prototyping virtual reality serious games for building earthquake preparedness: the Auckland city hospital case study, *Advanced Engineering Informatics*, 38, 670–682.

Lucas, J., Thabet, W. & Worlikar, P., (2007). Using virtual reality (VR) to improve conveyor belt safety in surface mining. *24th W78 Conference Maribor 2007 & 5th ITCEDU Workshop & 14th EG-ICE Workshop: Bringing ITC Knowledge to Work*, pp. 431–438.

Manjrekar, S., Sandilya, S., Bhosale, D., Kanchi, S., Pitkar, A., & Gondhalekar, M. (2014). Cave: An emerging immersive technology. *UKSim-AMSS 16th international conference on computer modelling and simulation* (pp. 131–136). IEEE.

Market analysis report (2020). *Virtual Reality Market Size, Share & Trends Analysis*, 2020 – 2027. Available from: https://www.grandviewresearch.com/industry-analysis/virtual-reality-vr-market . Accessed 30/12/2020

Matini, M. R., Andaroodi, E., & Ono, K. (2019). A 3D approach to reconstitution of the adobe citadel of Bam after earthquake: a complementary interpretation of architectural heritage knowledge, aerial photogrammetry, and heterogeneous data. *International Journal of Architectural Heritage*, 13(4), 600-618.

Meißner, M., Pfeiffer, J., Pfeiffer, T., & Oppewal, H. (2017). Combining virtual reality and mobile eye tracking to provide a naturalistic experimental environment for shopper research. *Journal of Business Research*, 100, 445-458.

Meng, F. & Zhang, W. (2014). Way-finding during a fire emergency: an experimental study in a virtual environment. *Ergonomics*, 57, 816-287

Merchant, Z., Goetz, E. T., Cifuentes, L., Keeney-Kennicutt, W., & Davis, T. J. (2014). Effectiveness of virtual reality-based instruction on students' learning outcomes in K-12 and higher education: A meta-analysis. *Computers & Education*, 70, 29–40.

Milgram, P., & Kishino, F. (1994). A taxonomy of mixed reality visual-displays, In: *1st International Working Networked Reality* (NR 94), Institute of Electrical and Electronics Engineers Inc. pp. 1321–1329.

Nishihara, A. (2015). Object recognition in assembly assisted by augmented reality system object recognition in assembly assisted by augmented reality system. *IEEE Xplore.*

Pan, Z., Cheok, A. D., Yang, H., Zhu, J., & Shi, J. (2006). Virtual reality and mixed reality for virtual learning environments. *Computers & Graphics*, 30(1), 20–28.

Pappas, N. & Glyptou, K. (2021). Accommodation decision-making during the COVID-19 pandemic: Complexity insights from Greece. *International Journal of Hospitality Management*. 93, 102767

Pappas, N. (2019) Crisis management communications for popular culture events. *Event Management*, 23(4-5), 655-667.

Parapanos, D., and Michopoulou, E. (2019). Understanding key motivations for using a hotel gamified application. In Pesonen, J., and Neidhardt, J. (eds.) *Information and Communication Technologies in Tourism 2019*, Vienna-New York: Springer, pp. 411- 422

Park, M. & Yoo, J. (2020). Effects of perceived interactivity of augmented reality on consumer responses: a mental imagery perspective. *Journal of Retailing and Consumer Services*. 52, 101912.

Pauchant, T. C., & Mitroff, I. I. (1992). *Transforming the Crisis-Prone Organization: Preventing individual, organizational, and environmental tragedies*. San Francisco: Jossey-Bass.

Peddie, J. (2017). *Augmented reality: Where we will all live*. Cham: Springer.

Perlman, A., Sacks, R. & Barak, R. (2014). Hazard recognition and risk perception in construction. *Safety Science*. 64, 13-21.

Pourmand, A., Davis, S., Lee, D., Barber, S. & Sikka, N. (2017). Emerging utility of virtual reality as a multidisciplinary tool in clinical medicine. *Games for Health Journal*. 6, (5)

Rauschnabel, P. A., Rossmann, A., & tom Dieck, M. C. (2017). An adoption framework for mobile augmented reality games: The case of Pokémon Go. *Computers in Human Behaviour*, 76, 276–286.

Revfine (2021). *How Augmented Reality is Revolutionising the Travel Industry*. https://www.revfine.com/augmented-reality travel-industry/ . Accessed: 15/01/2020

Rosenthal, U., & Kouzmin, A. (1993). Globalizing an agenda for contingencies and crisis management. *Journal of Contingencies and Crisis Management*, 1(1), 1-12.

Scarles, C., Casey, M., & Treharne, H. (2016). Enriching the visitor experience: Augmented reality and image recognition in tourism. In: M. Scerri, & L. K. Hui (Eds.). *The Changing Landscape of Tourism and Hospitality: The impact of emerging markets and emerging destinations* (pp. 1177–1184). Sydney: CAUTHE.

Shah, M. (2019) *How Augmented Reality (AR) is Changing the Travel & Tourism Industry*. Towardsdatascience. https://towardsdatascience.com/how-augmented-reality-ar-is-changing-the-travel-tourism-industry-239931f3120c . Accessed:26/12/2020

Sigala, M. (2018). New technologies in tourism: From multi-disciplinary to anti-disciplinary advances and trajectories. *Tourism Management Perspectives*, 25, 151–155.

Skare, M., Soriano, D.R. & Porada-Rochon.,M. (2021). Impact of COVID-19 on the travel and tourism industry. *Technological Forecasting & Social Change*. 163,120469.

Smith, S. P., & Trenholme, D. (2009). Rapid prototyping a virtual fire drill environment using computer game technology. *Fire Safety Journal*, 44, 559–569

Sönmez, S. F., Backman, S. J., & Allen, L. (1994). *Managing Tourism Crises: A guidebook*. Stafford: Clemson University.

Syal,S. & Mathew, R. (2020)Threats faced by mixed reality and countermeasures. *Procedia Computer Science*, 171 , 2720–2728

Tamura, H., Yamamoto, H., & Katayama, A. (2001). Mixed reality: Future dreams seen at the border between real and virtual worlds. *IEEE Computer Graphics and Applications*, 21(6), 64-70.

Thiessen, A. (2011). *Organisationskommunikation in Krisen – Reputationsmanagement durch situtative, integrierte und strategische Krisenkommunikation*. VS Verlag für Sozialwissenschaften.

Thomas, G., Gibson, J., Burlington,T. & Weir, B. (2019). *5G Smart Tourism Trial at the Roman Baths*. BBC. https://www.bbc.co.uk/rd/blog/2019-02-5g-mobile-augmented-reality-bath. Accessed:17/12/2020 tom Dieck, M. C., Jung, T., & Michopoulou, E. (2019). Experiencing virtual reality in heritage attractions: Perceptions of elderly users. In: M. C. tom Dieck, & T. Jung (Eds.), *Augmented Reality and Virtual Reality: The power of AR and VR for business* (pp. 89–98). Cham: Springer International Publishing.

Toubekis, G., Mayer, I., Doring-Williams, M., Maeda, K., Yamauchi, K., Tangiuchi, Y., Morimotod,S., Petzete,M., Jarkef, M .& Jansena,M.(2009). Preservation and management of the UNESCO world heritage site of Bamiyan: Laser scan documentation and virtual reconstruction of the destroyed Buddha figures and the archaeological remains. *22nd CIPA symposium,* Kyoto, Japan.

Tussyadiah, I. (2014). Expectation of travel experiences with wearable computing devices. In: Z. Xiang, & I. Tussyadiah (Eds.). *Information and Communication Technologies in Tourism* (pp. 539–552). New York: Springer International Publishing.

Tussyadiah, I. P., Wang, D., Jung, T. H., & tom Dieck, M. C. (2018a). Virtual reality, presence, and attitude change: Empirical evidence from tourism. *Tourism Management,* 66, 140–154.

Tussyadiah, I., Jung, T., & tom Dieck, M. C. (2018b). Embodiment of wearable augmented reality technology in tourism experiences. *Journal of Travel Research,* 57(5), 597–611.

Ukpabi, D. C., & Karjaluoto, H. (2016). Consumers' acceptance of information and communications technology in tourism. *Telematics and Informatics,* 34(5), 618–644.

Wei, W. (2019). Research progress on virtual reality (VR) and augmented reality (AR) in tourism and hospitality: A critical review of publications from 2000 to 2018. *Journal of Hospitality and Tourism Technology,* 10(4), 539–570.

Wiederhold, B.K. & Bouchard,S. (2014). *Advances in Virtual Reality and Anxiety Disorders,* Cham: Springer.

Zhang, T. (2021). Research on environmental landscape design based on virtual reality technology and deep learning. *Microprocessors and Microsystems.* 81, 103796

Zhou, J. (2021). Virtual reality sports auxiliary training system based on embedded system and computer technology. *Microprocessors and Microsystems.* 82,103944

Zhu, Y., & Li, N. (2021). Virtual and augmented reality technologies for emergency management in the built environments: A state-of-the-art review. *Journal of Safety Science and Resilience,* 2(1), 1-10.

Zyda, M. (2005). From visual simulation to virtual reality to games. IEEE *Comput. Xplore Digital Library,* 38(9), 25–32.

Part 3: Hospitality

Preface

This is the third part of the book consists of three chapters. Chapter 11 is co-authored by Nikolaos Pappas and Andreas Papatheodorou. The chapter examines complexity of attributes affecting the online purchasing process in tourism and hospitality, whilst it provides a comparison towards linear and non-linear analysis. In Chapter 12 Christina K. Dimitriou focuses on hospitality education and address the issues that academia has been struggling with, offering practical recommendations that will help tackle the operational difficulties of education. Chapter 13 is the last chapter of this book. It is authored by Maria Zoi Spanaki, and highlights the meaning of the risk management sector, and points out the importance of recognizing it as an essential field for the global economy.

11 Complexity in the tourism and hospitality online purchasing process

Nikolaos Pappas and Andreas Papatheodorou

Abstract

This chapter examines the complexity of attributes affecting the online purchasing process in tourism and hospitality. Eight elements (i.e. marketing activities, perceived benefits, perceived risks, over-choice confusion, price issues, quality issues, consumer trust and intention to purchase) are identified as of principal importance. Based on a sample of 735 holidaymakers, the empirical findings stress the role of perceived benefits in e-marketing activities; the significance of perceived risks for price and quality issues; and the importance of over-choice confusion to e-shopping. The research employs Qualitative Comparative Analysis (QCA), an innovative technique in tourism. QCA results are compared to regression highlighting the possible superiority of this non-linear mixed method against correlational analysis. The chapter progresses from fit to predictive validity, something that only a handful of studies have employed in tourism and hospitality, and generally in the service sector. It also discusses several theoretical, methodological and managerial implications, whilst it contributes in both, theoretical and methodological domains.

Keywords:

Online tourism shopping; chaos and complexity; benefits and risks; over-choice confusion; price and quality; e-consumer trust

Introduction

Information Technology has heavily impacted tourism and hospitality and the way travellers acquire and use relevant information (Xiang et al., 2015). Over 10% of Internet searches are related to travel (Jansen et al., 2008; Pappas, 2017), while 60% of people 15 years old or older tend to use the Web to search tourism and hospitality information and purchase relevant products (No & Kim, 2015). Not surprisingly, therefore, tourism and hospitality oriented web-vendors have become very active, significantly influencing the access of travellers to tourism and hospitality products and services (Xiang et al., 2008).

Hyun (2006) classifies tourist and hospitality information websites into portals, travel agencies, public institutions, online communities, and others. Based on the provision of information, Pan and Fesenmaier (2000) have divided tourism and hospitality related e-vendors into travel and tourism websites. From a consumer's point of view, the beneficial impact of Internet shopping predominantly relates to time and place convenience; value enhancement thanks to effective price comparison; and hedonic consumption arising from the very purchasing experience (Sigala, 2004). On the other hand, a number of Internet use caveats have been identified such as privacy and security concerns (Evans & Mathur, 2005); online versus offline risks (Hong & Yi, 2012); price and quality issues (Pappas, 2017); and over-choice confusion (Bjork & Kauppinen-Raisanen, 2012). Still, research on the complexity of online tourism and hospitality purchasing decisions and behaviour is rather limited. This is surprising given the role of technology in shaping consumer behaviour in advanced modern societies (Chai, 2012).

From a methodological point of view, most business-oriented and almost all tourism and hospitality studies evaluate statistical relationships based on symmetric analysis as encapsulated by structural equation modelling (SEM) and regression. These linear methods adopt a net effect estimation approach but ignore the complexities that exist in reality and are apparent in academic datasets (Woodside, 2014). This is because when multicollinearity is high, estimates may prove statistically insignificant; alternatively, estimates can prove inconsistent with the set hypotheses since the estimated regression function is of poor predictive power (Van der Meer et al., 2005). Conversely, in cases of low multi-collinearity the marginal contribution of one explanatory variable may end up depending on the other explanatory variables included in the estimated function albeit in a non-linear manner (Woodside, 2013). In fact, the standard assumption in regression analysis is that the addition of new variables increases (at least the adjusted) goodness-to-fit (Armstrong, 2012); nonetheless, the usual co-variance predictors in non-experimental studies do not provide any related supportive evidence (Skarmeas et al., 2014).

This chapter focuses on website vendors in tourism and hospitality, whilst its full linear analysis is presented in Pappas (2017). It synthesises previous research (i.e. what we know) on the main elements (variables) affecting tourism online purchasing intentions in terms of marketing activities (Chikweche & Fletcher, 2010); perceived benefits (Kim et al., 2008); risks (Hong & Yi, 2012) and confusion by over-choice (Tarnanidis et al., Omar, 2015); price (Tarnanidis et al., 2015) and quality issues (Ahn, Ryu, and Han, 2004); consumer trust (Gefen et al., 2003); and intention to purchase (Kim et al., 2008). Most importantly, however, it progresses from the parametric analysis illustrated in Pappas (2017) to a non-linear evaluation of the examined aspects, as it steps beyond a simple synthesis to discuss what we do not know very well, i.e. how complexity affects the attributes of tourism and hospitality online purchasing intentions. Given the sheer scale of online bookings and the inherent risks associated with Internet use, understanding and evaluating consumer behaviour through the theoretical lens of complexity is a relatively uncharted territory worth exploring further. To do so, the paper implements Qualitative Comparative Analysis (QCA), which is relatively new in tourism and hospitality research, while only a few studies have generally employed it in the service sector (Papatheodorou & Pappas, 2017). On these grounds, the contribution of the study is both theoretical and methodological. Concerning literature, it provides a better understanding and possible explanation of online tourism and hospitality decisions based on complexity theory. Further to the study of Pappas (2017), it suggests multiple pathways leading to the same outcome, instead of solely based on only one. As a result, it also provides a series of managerial implications addressing the complexity of online tourism behaviour. In terms of methodology, it applies an innovative asymmetric analysis technique whose possible superiority and suitability for tourism research is assessed by directly comparing its results with dominant linear (i.e.

Structural Equation Modelling) regression analysis using Pappas (2017) and its associated dataset as a benchmark. The paper also steps forward from fit to predictive validity for the proposed models.

Chaos and complexity

Chaos theory suggests that even small changes can produce substantial diverging outcomes to dynamic systems making it impossible to predict behavioural patterns in the longer term (Kellert, 1993). Complexity theory emerged from chaos theory and focuses on systems with complex characteristics, i.e. it "*...deals with systems that have many interacting agents and although hard to predict, these systems have structure and permit improvement*" (Zahra & Ryan, 2007: 855). Both theories work with nonlinear systems and are sensitive to initial conditions; nonetheless, chaos theory leaves no room for forecasting, while in complexity theory unpredictable behaviour is set and analysed within a quasi-stable context (Olmedo & Mateos, 2015).

More specifically, in the theory of chaos simple systems generate complicated non-predictable behavioural patterns, while the theory of complexity focuses on the way multi-elemental systems lead to relative behavioural predictability (Baggio, 2008). Even if the complexity of chaordic outcomes may result in dramatic and unpredictable situations the emerging dynamic systems are not totally uncontrolled, as some kind of order still exists (Nilson, 1995; Zahra & Ryan, 2007). Thus, in social sciences, complexity theory and Qualitative Comparative Analysis (to be discussed later) may be used to explain consumer evaluations, attributes, and the process of decision-making by implementing alternative asymmetric (nonlinear) combinations of various indicators (Wu et al., 2014).

The tourism and hospitality perspective

Tourism and hospitality research has not adequately focused on complexity so far, as reductionist approaches still prevail (McDonald, 2009). In reality, however, the behavioural patterns of travellers depend on numerous factors and produce complex relationships with inherent nonlinearity, which prevents the emergence of direct causal links (Olmedo & Mateos, 2015). In fact, tourist behaviour is likely to be affected by both endogenous and exogenous systemic factors (Boukas & Ziakas, 2014;

Dylan, 2017); even so, all tourism and hospitality related factors have emerging stable features as some order seems to exist in their configuration (Olmedo & Mateos, 2015). This is the reason why almost two decades ago, chaos and complexity theories were proposed by Faulkner and Russell (1997) as alternative frameworks to examine tourism attributes and better understand dynamic systems whose comprehension until then was poor, due to the assumption that all underlying relationships are stable and static. In other words, the extent of behavioural complexity makes the Newtonian (linear) thinking inadequate and indicates a need for asymmetric (nonlinear) analysis (Laws & Prideaux, 2005). In tourism and hospitality, the application of complexity theory can provide substantial information concerning the formulation and expression of behavioural patterns (Russell & Faulkner, 2004), helping to better comprehend the changing dynamics of the tourist system (Faulkner & Russell, 2000).

Background of the study

The importance of Internet as a distribution channel in tourism and hospitality is continuously growing, and its revenue generation worldwide is expected to reach 523 billion US dollars in 2016 (Statista, 2016). In fact, e-commerce has rapidly grown mainly due to the convenience and the value for money it offers to customers and the ability for them to learn about firms, products and services (Marom & Seidmann, 2011). On the other hand, perceived risk is included in all shopping decisions, predominantly when the outcome is uncertain (Dholakia, 2001). Hence, e-commerce has created concerns especially with respect to privacy (Pantano et al., 2013), security (Taylor & Strutton 2010), time consumed searching for information; uncertainty dealing with after sales service warranty when compared with traditional ways of shopping (Hong & Yi, 2012) and general distrust in product quality and delivery (Hong & Cha, 2013). Nonetheless, tourism and hospitality research related to retail online shopping behaviour is somewhat limited (Law et al., 2009; Amaro & Durate, 2015).

Marketing activities refer to the development, distribution, pricing and promotion of products and services (Dibb & Simkin, 2013; Pappas, 2015). Marketing can substantially influence the beliefs and perspectives of consumers about the performance of products and services (Nerkar & Roberts 2004) and ultimately determine their likelihood to purchase

(Leenders & Wierenga, 2008). Online marketing and eWOM is an essential tool for the minimisation of perceived risks introduced by frequent technological advances; the rapid technological evolution and development amongst consumers and competitors; and the reduction of overchoice confusion due to the increasing volume of product and service alternatives available to customers (Bore et al. 2017; Pantano et al., 2013). On the other hand, the benefits and risks of online shopping and the trust of consumers in products affect the e-marketing process (Pescher et al., 2014), since this needs to consider the characteristics of consumers (purposive and entertainment values); the provision and quantity of alternative products; and the formulation process of perceived benefits and risks (Okazaki, 2008).

In particular, the perceived e-commerce risks negatively affect consumers (Kim, 2007) irrespectively of how experienced shoppers they are (Liang & Jin-Shiang, 1998). This negative relationship increases when the risks are associated with monetary losses (Keating et al., 2009). These may prove relatively substantial, e.g. in the case of a family holiday package. In addition, tourists may be overloaded by the massive amounts of information provided by tourism and hospitality oriented web-vendors, as this may then negatively affect the image of tourist products, services, and destinations (Bjork & Kauppinen-Raisanen, 2012). Moreover, the high extent of specialisation and/or similarities of the provided products and services in tourism and hospitality increases the customers' confusion and renders purchasing decisions even more difficult (Yang and Lai, 2006).

Price and quality also affect online shopping. E-vendors provide price comparisons, giving the ability to consumers to find the lowest price for each product and service they are interested in buying (Bruce et al., 2004). The ease of comparing prices and finding product level information through online shopping is unprecedented, and creates new challenges for effective price discrimination (Garbarino & Maxwell, 2010). Furthermore, the importance of web-vendor quality leads companies to improve their websites' design, and enhance the interaction experience of their customers (Kholoud Al-Qeisi et al., 2014). The quality of vendors is a fundamental criterion for the formulation of purchasing and repurchasing intentions over an online channel (Bhatnagar et al., 2003). Thus, when an online channel is characterised by high quality, the consumers tend to trust it, hence it increases the possibility to be used by potential clients (Montoya-Weiss et al. 2003). Moreover, the information level (quality, accuracy etc.) provided by the web-vendor, and the consumers' perceptions in terms of product quality this vendor offers, also influence the relationship between price and quality (Ahn et al., 2004).

Having the above in mind, it is now widely accepted that Information Technology has changed the way consumers make their purchasing decisions (Moran & Muzellec, 2017). E-retailers substantially emphasise consumer trust, since e-shoppers are more reluctant to purchase the products in which they are interested online (Park et al., 2012). This is due to the inability of consumers to visit a store with physical presence and examine in person the product they intend to buy (Hong & Cho, 2011). Cognitive trust, which focuses on the beliefs of consumers stemming from logical expectations about the attributes of online retailers, impacts on emotional trust, which concerns the attitudes of consumers and their emotions, and this further impacts upon the intention to purchase (Komiak & Bembasat, 2006). In addition, the level of trust in e-shoppers exposed to inconsistent information and revisions in terms of products and services considerably influences their purchasing intentions (Zhang et al., 2014). This is the reason why e-retailers should try to convince consumers to buy their products (purchase decision and money transfer), by exceeding the trustworthiness behaviour threshold (Bente et al., 2012).

Methods

Participants

The empirical research undertaken in this paper focuses on holidaymakers returning to Manchester International Airport (IATA code: MAN) who had used the Internet to book a part (travel, accommodation, destination activities) or the whole range of their holiday activities. This study used structured personal interviews with structured questionnaires as the most appropriate method to obtain the primary data. Personal interviews were the best method of achieving the study's objectives since they are the most versatile and productive method of communication (Pappas, 2014). They facilitate spontaneity and also provide opportunities to guide the discussion back to the outlined topic when discussions prove unfruitful (Sekaran & Bougie 2009). The participants' selection was based on an exclusion question at the beginning of the interview which asked whether the potential participants had purchased online services related to their current vacations. More specifically, a stratified sample was selected by interviewing travellers at the airport's train station; bus station; and parking facilities. Following Sekaran and Bougie's (2009) suggestions for bias minimisation, one out of three travellers approaching the targeted areas was asked to participate in the research. Potential respondents were briefed on the study purpose; informed about the voluntary character of their participation; and assured of full confidentiality and anonymity of responses. Although the level of missing data vis-à-vis the entire sample was very small (i.e. only eight respondents declined to complete the interview), list-wise deletion was used (leading to the exclusion of the specific interview from the analysis) since this is considered as the least problematic method of handling missing data (Allison, 2001).

Sample determination and collection

Representativeness is a fundamental requirement when determining the sample size. According to Sevgin et al. (1996), when there are unknown population proportions, the researcher should choose a conservative response format of 50/50, based on the assumption that 50% of the respondents have negative perceptions, and 50% have not, to determine the sample size. A confidence level of at least 95% and a 5% sampling error were selected. Based on the above, the sample size was:

$$N = \frac{(t - table)^2 (hypothesis)}{S^2} \rightarrow N = \frac{(1.96)^2 (0.5)(0.5)}{(0.5)^2} \Rightarrow N = 384.16 \ [Rounded\ to\ 400]$$

The calculation of the sampling size is independent of the total population size, hence the sampling size determines the error (Aaker & Day 1990). Participants were approached in MAN's train station (400 people), bus station (400 people), and car parking facilities (400 people). Of the 1,200 holidaymakers approached, 735 completed the questionnaire (response rate: 61.25%). The overall statistical error of the sample population was 3.6%.

11

Measures

The questionnaire was based on prior research (Pappas, 2017) and consisted of 39 statements (Table 11.1) rated using a Likert Scale (1 strongly agree/7 strongly disagree), plus one exclusion question concerning online purchasing of tourist/hospitality products.

Table 11.1: Descriptive statistics. Source: adapted from Pappas (2017)

Marketing Activities		Mean	St. Dev
MA1	Direct marketing activities (i.e. direct mails and e-mails) influence my purchasing online decisions	2.29	0.56
MA2	The "above the line" promotional activities (i.e. TV and radio advertisements) influence my purchasing online decisions	3.02	0.57
MA3	Corporate branding strategy influences my purchasing online decisions	2.78	0.47
MA4	The performance of the product I intend to buy influences my purchasing online decisions	1.75	0.26
Perceived Benefits		Mean	St. Dev
PB1	I believe that online shopping is convenient	2.16	0.60
PB2	I can save money by shopping online	1.87	0.74
PB3	I can save time by shopping online	1.55	0.57
PB4	Purchasing online enables me to accomplish a shopping task more efficiently than using high street stores	2.47	0.37
PB5	Purchasing online increases my shopping productivity	2.83	0.49
Perceived Risks		Mean	St. Dev
PR1	Purchasing online involves greater risk of credit loss compared with traditional shopping	2.55	0.78
PR2	Purchasing online involves a more complicated payment procedure compared with traditional shopping	2.21	0.36
PR3	Purchasing online involves more time spent searching for information compared with traditional shopping	5.28	0.45
PR4	Purchasing online involves greater risk of private information loss compared with traditional shopping	4.06	0.70
PR5	Purchasing online involves greater risk of after-sales service warranty complications compared with traditional shopping	3.87	0.44
PR6	Providing credit card information online is riskier compared to doing so over the phone to an offline vendor	4.85	0.48
PR7	Purchasing online involves a risk of fraudulent website practices	1.88	0.71
Over-Choice Confusion		Mean	St. Dev
CO1	There are so many tourism products to choose from that often I feel confused	2.19	0.48
CO2	Sometimes it is hard to choose from where to shop	3.45	0.47
CO3	The more I learn about tourism products, the harder it seems to choose the best	4.20	0.37
CO4	All the information I get on different tourism products confuses me	2.74	0.55
Price Issues		Mean	St. Dev
PI1	The higher the price of a product, the better its quality	3.01	0.45
PI2	I prefer to buy the best-selling brands	2.82	0.65
PI3	I buy as many of my tourist products as possible at discounted prices	3.68	0.88
PI4	I usually choose lower priced tourist products	1.42	0.72
PI5	I look carefully to find the best value-for-money	1.78	0.53
Quality Issues		Mean	St. Dev
QI1	It is important that the e-vendor provides detailed information on featured products	1.82	0.54
QI2	It is important that the e-vendor provides accurate information on featured products	1.97	0.60
QI3	It is important that the e-vendor can be trusted to provide whatever is promised	1.70	0.85
QI4	It is important that the e-vendor instils confidence in customers by reducing uncertainty (e.g. joint problem-solving)	1.52	0.74
QI5	It is important that the e-vendor understands and adapts to the customer's specific needs	1.69	0.62
QI6	It is important that the e-vendor sells tourism products of high quality	2.46	0.41
QI7	It is important that the e-vendor offers a wide variety of tourism products	2.88	0.46
Consumer Trust		Mean	St. Dev
CT1	Shopping online is trustworthy	2.95	0.69
CT2	The e-vendor I use gives me the impression of being honest	2.87	0.55
CT3	The e-vendor I use gives me the impression that cares for their customers	2.41	0.49
CT4	The e-vendor I use gives me the impression that they have the ability to fulfil my needs	2.56	0.78
Intention to Purchase		Mean	St. Dev
IP1	I am likely to purchase tourism products online	2.24	0.51
IP2	I am likely to recommend online shopping to my friends	1.90	0.59
IP3	I am likely to make another online purchase if I need a tourism product	2.07	0.47

The reliability and validity of this selection rationale is supported by studies such as Kyle et al. (2003) and Gross and Brown (2008). Eight statements were selected from six different studies, i.e. Chikweche and Fletcher (2010) with respect to marketing activities (MA – four statements); Kim et al. (2008) to evaluate the perceived benefits (PB – five statements) and the intention to purchase (IP – three statements); Hong and Yi (2012) to examine the perceived risks (PR – seven statements); Tarnanidis et al. (2015) regarding over-choice confusion (CO – four statements) and price issues (PI – five statements); Ahn et al. (2004) to address quality issues (QI – seven statements); and Gefen et al. (2003) with respect to consumer trust (CT – four statements).

fsQCA

The study analyses the participants' responses to the questionnaire using fuzzy-set Qualitative Comparative Analysis (fsQCA). This is a theoretical method to examine relationships believed to have a bearing upon the outcome of interest and to identify any potential binary set of combinations generated from related predictors (Longest & Vaisey, 2008). QCA is a mixed-method technique, since it combines quantitative empirical testing (Longest & Vaisey, 2008) and qualitative inductive reasoning through case analysis (Ragin, 2000). QCA handles logical complexity by allowing for the fact that different combinations of characteristics may produce different results when jointly considered with other events or conditions (Kent & Argouslidis, 2005). The study also estimates negated sets, i.e. presence or absence of a given condition (Woodside & Zhang, 2013), where membership is calculated by taking one minus the score of membership of the examined case in the original fuzzy set (Skarmeas et al., 2014).

According to Ordanini et al. (2014), in set theory a sub relation with fuzzy measures is consistent when the membership scores of a given attributional causal set are equal or consistently less than the membership scores in the outcome set. Thus, consistency should be calculated as follows:

11

$$Consistency\ (X_i \leq Y_i) = \sum_i [\min(X_i; Y_i)] \ / \sum_i (X_i)$$

where, for holidaymaker i, X_i is the membership score in each considered attribute (i.e. configuration) and Y_i is the membership score in the outcome condition. Accordingly, coverage includes the assessment of sufficient configurations' empirical importance (Ordanini et al., 2014) and is calculated as follows:

$$Coverage\ (X_i \leq Y_i) = \sum_i[\min(X_i;Y_i)] \,/\, \sum_i(Y_i)$$

As presented in the correlation matrix (Table 11.2) the absolute values of all correlation coefficients are below 0.60.

Table 11.2: Correlation matrix

Measures		1	2	3	4	5	6	7	8
1	Marketing Activities	1.00							
2	Perceived Benefits	.31	1.00						
3	Perceived Risks	-.26	-.06	1.00					
4	Over-choice Confusion	-.28	.03	.17	1.00				
5	Price Issues	-.17	-.04	.25	.27	1.00			
6	Quality Issues	.08	.03	-.13	-.18	.12	1.00		
7	Consumer Trust	.09	.01	-.08	-.32	.07	.25	1.00	
8	Intention to Purchase	.22	.12	-.04	-.11	.02	.31	.12	1.00
	Cronbach α	.831	.828	.837	.809	.846	.848	.837	.829

According to Skarmeas et al. (2014), this outcome indicates that there is a general asymmetry in the various relationships among variables. Thus, the causal conditions produced by the alternative combinations may lead to the same outcome condition (Woodside, 2013). Using fsQCA this study examines the formulation of online purchasing tourism and hospitality intentions by estimating the complex antecedent conditions (i.e. causal recipes) leading to high membership in the eight statements discussed previously. Figure 11.1 shows graphically the underlying research model and the relationships expected to emerge among the statements. The theoretical justification of the model constructs as well as the determination of its hypotheses (with the full deployment of its linear analysis), are presented in Pappas (2017).

While traditional analysis considers arrows in the context of linearity, in the present case non-linear complex solutions (i.e. causal recipes) are of importance as depicted by the level of membership scores. In fact, as Woodside and Zhang (2013) indicate, the membership score of a case in a recipe is the membership degree of simple causal conditions of fuzzy-sets that intersect and include the recipe. In the causal recipe, this intersection is the minimum score amongst the selected simple conditions (Skarmeas et al., 2014). Based on complexity, the study assumes that instead of having symmetrical net effects, asymmetric relationships exist.

According to Woodside (2014: 2499) the asymmetric measure of consistency is analogous to the symmetric measure of correlation, whilst the asymmetric measure of coverage is analogous to the symmetric coefficient of determination.

When examining model acceptability, the provided solution is considered informative when coverage is between .25 and .75 and consistency is above .74 (Skarmeas et al., 2014).

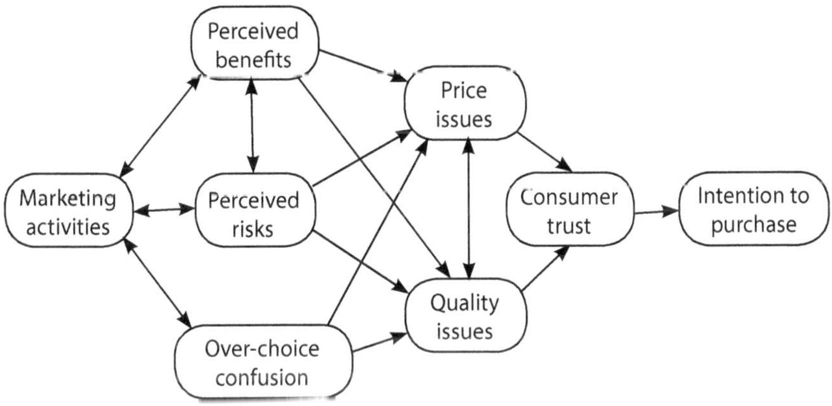

Figure 11.1: The Research Model Architecture. Adapted from Pappas (2017)

Results

Implementation of fsQCA algorithms

Having the above in mind, the study evaluates the causal recipes (or complex antecedent conditions) that provide a high membership. Using non-linearity, the study highlights combined complexities and identifies asymmetric relationships. The research includes the perspectives of 735 respondents, and its calibration has been implemented by a group of 47 randomly selected individual cases. The calibrated fuzzy sets used were 'f_ma' for marketing activities; 'f_pb' for perceived benefits; 'f_pr' for perceived risks; 'f_oc' for over-choice confusion; 'f_pi' for price issues; 'f_qi' for quality issues; 'f_ct' for consumer trust; and 'f_ip' for intention to purchase.

11

Table 11.3: Complex solutions

Complex Solution Findings	Raw Coverage	Unique Coverage	Consistency
Marketing Activities			
Model: f_ma=f(f_pb,f_pr,f_oc)			
f_pb*f_pr*~f_oc	0.429	0.076	0.869
f_pb*~f_pr*~f_oc	0.396	0.113	0.873
Solution Coverage: 0.414			
Solution Consistency: 0.833			
Perceived Benefits			
Model: f_pb=f(f_ma,f_pr)			
f_ma*f_pr	0.482	0.483	0.849
Solution Coverage: 0.482			
Solution Consistency: 0.850			
Perceived Risks			
Model: f_pr=f(f_ma,f_pb)			
f_ma*~f_pb	0.521	0.528	0.859
Solution Coverage: 0.521			
Solution Consistency: 0.859			
Price Issues			
Model: f_pi=f(f_pb,f_pr,f_oc,f_qi)			
f_pb*f_pr*~f_oc*f_qi	0.403	0.089	0.830
f_pb*f_pr*f_oc*~f_qi	0.328	0.079	0.818
Solution Coverage: 0.383			
Solution Consistency: 0.814			
Quality Issues			
Model: f_qi=f(f_pb,f_pr,f_oc,f_pi)			
f_pb*f_pr*f_oc*~f_pi	0.393	0.089	0.814
~f_pb*f_pr*~f_oc*f_pi	0.439	0.113	0.820
f_pb*f_pr*~f_oc*f_pi	0.372	0.083	0.809
Solution Coverage: 0.517			
Solution Consistency: 0.815			
Consumer Trust			
Model: f_ct=f(f_pi,f_qi)			
f_pi*f_qi	0.508	0.508	0.884
Solution Coverage: 0.508			
Solution Consistency: 0.884			

Sufficient complex statements for marketing activities

In terms of marketing activities, the complex solution reveals two pathways (Table 11.3). The first concerns the combination of high influence of perceived benefits and risks with low influence (denoted by ~) from over-choice confusion. The consistency of this pathway is good (0.869), explaining a considerable amount of marketing activities (coverage=0.429). The second solution reveals that high perceived benefits with low perceived risks and low over-choice confusion lead to high membership scores for marketing activities. This pathway is slightly more consistent than the previous one (consistency=0.873), but explains fewer cases of marketing activities (coverage=0.396). In total, the marketing activities solution indicates a fairly high consistency (0.833) and coverage (0.414).

Sufficient complex statements for perceived benefits and risks

As illustrated in Table 11.3, the complex solution revealed only one pathway concerning perceived benefits. This pathway suggests that high marketing activities and perceived risks lead to high membership scores for perceived benefits. The consistency of this pathway is good (0.850) also having a considerable inclusion of cases (coverage=0.482).

As in perceived benefits, the complex solution revealed only one pathway in perceived risks. More specifically, the results indicate that high marketing activities and low perceived benefits produce high membership scores for perceived risks. This solution is also characterised by fairly good consistency (0.859) and the highest coverage (0.521) produced by any other complex solution.

Sufficient complex statements for price and quality issues

There are two complex solutions produced in terms of price issues. The first solution proposes that high perceived benefits and risks, and high quality issues with low over-choice confusion result in high price issues. This solution is fairly consistent at 0.830 and has a good coverage at 0.403. The second solution suggests that high perceived benefits, risks and over-choice confusion with low quality issues also result in high price issues. This complex solution gives a slightly lower consistency (0.818) than the previous one, also having a lower coverage (0.328). The solution as a whole has a good consistency at 0.814 and high coverage at 0.383.

Quality issues seem to generate most of the complex solutions. The fist solution suggests that high perceived benefits, risks and over-choice confusion with low price issues can produce high quality issues. The consistency of this solution is at 0.814 and its coverage at 0.393. The second solution indicates that high perceived risks and price issues with low perceived benefits and over-choice confusion lead to high membership scores for quality issues. This solution presents the highest consistency (0.820) and coverage (0.439) of the three generated in quality issues. The last solution suggests that high perceived benefits, risks and price issues with low over-choice confusion result in high quality issues. In contrast with the previous solution, this one generates the lowest consistency (0.809) and coverage (0.372) levels. Overall, the solution consistency is good (0.815) also providing a high coverage (0.517).

Sufficient complex statement for consumer trust

The findings provide one more complex solution, this time focusing on consumer trust. The last complex solution indicates that high price and quality issues provide high membership scores in consumer trust. This pathway appears to have the highest consistency (0.884) from all other complex solutions in this research. It also has a high coverage rate (0.508).

Based on the model architecture of Figure 11.1 and since confusion by over-choice and intention to purchase are solely determined by one factor (i.e. marketing activities and consumer trust respectively) no sufficient complex statements have been estimated in these cases.

Discussion

The results presented above set the foundations for an interesting discussion. In terms of marketing activities, perceived benefits are included in both complex solutions highlighting their importance for the consumers' online decision-making in tourism and hospitality. This finding is in agreement with Pescher et al. (2014) and Leenders and Wierenga (2008). On the other hand, perceived risks do not appear in both derived solutions, i.e. perceived risks do not always influence marketing activities related to tourism products. In addition, over-choice confusion is absent from both derived recipes. The two latter findings contradict Pantano et al (2013) and suggest that the fundamental relationship encapsulates marketing activities and perceived benefits, whilst in some cases perceived risks are also included in this complex solution. They also highlight how marketing activities are influenced by the three fundamental aspects (i.e.

perceived benefits, risks, and over-choice confusion) that subsequently impact on price, quality and consumer trust and ultimately affect purchasing intentions of online tourism consumers as shown in Figure 11.1.

Concerning perceived benefits, the combination of both antecedent conditions (i.e. marketing activities and perceived risks) is necessary to shape their influence in online tourism and hospitality shopping. This pathway is in line with Marom and Seidmann (2011) considering the beneficial effects of online shopping, and Dholakia (2001) regarding the inclusion of perceived risk in all purchasing decisions. On the other hand, the provided solution for perceived risks indicates their association with marketing activities, while perceived benefits are excluded from the derived recipe. This finding further highlights the importance of perceived risks in decision making as suggested by Dholakia (2001) and the influential mitigating role of marketing activities as pinpointed by Nerkar and Roberts (2004). These results contribute to the better understanding of perceived risks' role and importance for online tourism and hospitality shopping, and further clarify their relationship and influence on perceived benefits and marketing.

Price issues findings reveal that perceived benefits and risks appear in both pathways, making them the necessary antecedent conditions for price-oriented perception formulation. Nonetheless, when over-choice confusion appears (first recipe), quality issues are excluded and vice versa (second recipe). The findings indicate that all three fundamental aspects (i.e. perceived benefits, risks, and over-choice confusion) can influence price issues, while in some cases over-choice confusion does not affect purchasing decisions. Similarly, findings in quality issues have revealed three pathways. The only antecedent simple condition that appears in all derived recipes is the perceived risks. Conversely, perceived benefits and price issues appear in two out of the three solutions, while over-choice confusion is included only in one of them. This makes the latter condition the weaker one regarding its influence on quality issues compared to the other three. The significance of perceived risks indicates their importance for quality perceptions. In addition, when all three fundamental aspects (perceived benefits, risks, and over-choice confusion), impact upon quality, the price issues are absent. Thus, the relationship affecting price and quality as discussed by Ahn et al. (2004), and theoretically supported by the price-quality nexus, which suggests that customers use price to evaluate overall product superiority or excellence (Lichtenstein et al., 1993; Zeithaml, 1988), is not always confirmed by the present research.

The quality issues findings also reveal a mutually exclusionary role of over-choice confusion and price issues. This result contradicts Pappas (2017), who suggests that over-choice confusion has a direct significance for the formulation of quality issues, and affects both price and quality aspects. As the pathways indicate, the influence of the three fundamental factors and price issues upon quality issues is dependent on their combination and the extent they are included in decision-making, except for perceived risks that play an important role in all derived recipes.

Finally, the produced solution concerning consumer trust includes the combination of both price and quality antecedent conditions. This result confirms previous studies such as Bente et al. (2012), Pappas (2017), and Zhang et al. (2014). This recipe also has the highest consistency that appears in this research, showing the importance of these two factors in building consumer trust, and ultimately affecting customer intentions to purchase online tourism products.

QCA versus regression

As mentioned, most tourism studies evaluate statistical relationships from a linear perspective predominantly using structural equation modelling (SEM) and regression analysis. Thus, this paper implements additional analysis to compare QCA research findings with regression as extensively discussed in Pappas (2017). As Ordanini et al. (2014) suggest, any comparison should be made with caution since QCA implements distinct assumptions like complex causality; establishes relations by using cases instead of variables; focuses on different research objectives; and identifies configurations that provide sufficient and necessary conditions for a result of interest. If the provided comparison is not carefully implemented these aspects are possible to lead in meaningless outcomes.

In a nutshell and as discussed in Pappas (2017), the linear analysis shows that the Kaiser-Meyer-Olkin of Sampling Adequacy measure is 0.863, and the overall reliability measured through Cronbach α is .816, while all variables score over .8. All effects are statistically significant, while the overall model's R^2 is .521, and all expected associations are confirmed. More specifically, marketing activities have a positive interactive relationship with perceived benefits (β=.422; p<.01), and negative interdependence with perceived risks (β=-.357; *p*<.01) and over-choice confusion (β=-.186; *p*<.01). Perceived benefits positively affect prices (β=.314; *p*<.01) and quality issues (β=.280; *p*<.05), whilst the influence of perceived risks on price (β=-.327; *p*<.01) and quality issues (β=-.206; *p*<.05) is negative, as are the impacts of over-choice confusion (β=-.185; *p*<.05 and β=-.231; *p*<.01). In addition, perceived benefits and risks influence each other (β=.215; *p*<.05). Consumer trust is directly affected by price (β=.385; *p*<.01) and quality issues (β=.408; *p*<.01), while price and quality also affect each other (β=.219; *p*<.05). Finally, the linear analysis confirms the important positive impact of consumer trust on the intention to purchase (β=.411; *p*<.01). Comparing the results of the regression analysis with those from the QCA the appropriateness of the latter is clear. More specifically, the study of Pappas (2017) employs regression, which limits itself to the consideration of a single pathway, i.e. the joint linear direct effect of perceived benefits, risks and over-choice confusion on price and quality issues as well as on the interrelationship between price and quality. This indicates the inability of regression to encapsulate the full range of alternative combinations and influences that are capable of producing the same outcome, which is actually an inherent feature of complexity concerning decision-making formulation. For example, as highlighted in Table 11.3, the first fsQCA solution in price issues does not involve over-choice confusion (f_pb*f_pr*~f_oc*f_qi), whilst the second one does not include quality issues (f_pb*f_pr*f_oc*~f_qi), when, on the other hand, both are required by SEM. Furthermore, it is the fsQCA's sufficient configurations that exclude these conditions (one per solution), whilst in the study of Pappas (2017) SEM analysis indicates that the online decision-making process appears to be considerably dependent on those two aspects (over-choice confusion: β=.185/*p*<.05; qualitgy issues: β=.219/ *p*<.05). The comparison reveals that regression analysis can only partially explain the relationships between the examined constructs; on the other hand, QCA identifies complex relationships that cannot be encapsulated by the linear regression analysis.

Fit and predictive validity

Most studies focusing on the evaluation of specific models examine the model fit (Gigerenzer & Brighton, 2009) to ensure that the data support the relationships amongst the observed variables and their respective factors (Pappas, 2017). Thus, only a few studies focus on predictive validity

(Roberts & Pashler, 2000; Wu et al., 2014), since a good fit to observations does not necessarily indicate the existence of a good model (Gigerenzer & Brighton, 2009). The model's predictive validity ensures the configural model's predictive ability using a separate sample (Olya et al., 2018). This study also focuses on the estimation of the predictive validity. To test it, the process discussed by Wu et al. (2014) is followed: the research sample is divided in a holdout and a modelling subsample which is half the overall sample, since the patterns of job vulnerability are perceived as consistent indicators to produce high scores. The overall consistency exceeds .8 (C1=.823) and the coverage exceeds .5 (C2=.518). The results indicate that the model has good predictive validity.

Managerial challenges and applications

Regarding managerial aspects, the most important implication concerns the marketing strategies of tourism and hospitality e-retailers. As also emphasised by Nerkar and Roberts (2004) marketing activities can significantly influence and alter consumer perspectives in terms of products' and services' performance. As a new challenge, companies should clearly emphasise the promotion of tourism online product benefits, and strengthen their branding through direct online marketing. This may include the distribution of information via personal e-mails to potential or previous customers in terms of new products and services, optimisation of existing products, and forthcoming offers and events. This promotional activity can also include aspects of risk reduction combined with the beneficial impacts of online tourism and hospitality shopping. Another challenge could be the provision of product and e-vendor comparison of information and characteristics with other similar products and e-vendors existing in the market.

While in some cases the present research indicates gaps in the importance of price and quality, their significance as determinant factors remains strong. As a result, managers should closely examine their consumers' characteristics, regularly monitoring the formulation and alteration of their perceptions and preferences in the selection of both products and e-vendors. As indicated by the findings, the cornerstone for online purchasing remains the trust of consumers in e-vendors and products. Thus, it is imperative for enterprises engaged in online sales to provide specific services that reduce consumer uncertainties. Due to the intangibility of tourist and hospitality products these could include ad hoc information about destinations and products, post-purchase services, quality guarantees, accommodation characteristics etc. In this way, e-retailers will become able to better accommodate their customers, instil them with confidence and trust, and show genuine care for them.

Conclusion

This study has used fsQCA to examine complexity in online tourism and hospitality shopping. Italso provides a comparison of fsQCA use with regression, which is the dominant linear analysis method adopted in the tourism and hospitality sector, pinpointing the efficiency of the former in dealing with complex attributes, since it analyses cases instead of variables. It also demonstrates its predictive validity, something that only a handful of service related studies have done so far (Roberts & Pashler, 2000), highlighting the possible superiority of the provided models.

Despite the study's theoretical and methodological contribution, several limitations need to be highlighted. Since the implementation of fsQCA in tourism and hospitality is relatively new, the full potential of fsQCA is not yet realised, as this can only be assessed by implementing it in multiple tourism contexts involving chaos and complexity theories. A second aspect derives from the evaluation of different attributes, since this can produce different outcomes. Thus, it is advisable to implement the research with caution if other influential factors of tourism and hospitality (e.g. consumer familiarity with online bookings; the level of sophistication of the online bookings environment, and the e-vendor's platform ease of use) are to be examined. Third, the current study did not classify the informants in specific categories (e.g. according to socio-demographic characteristics), since its aim was to holistically examine the attributes affecting tourism online purchasing intentions, and introduce fsQCA as an innovative methodology in tourism, thus fruitfully capitalising on the extensive linear analysis discussed in Pappas (2017). While such a classification can be implemented at different model stages (as shown in Figure 11.1), its present application would immensely complicate the comparison of linear with asymmetric analysis thus making it more difficult to meet an important objective of this paper. Therefore, a comparative research between online and offline consumers or tourists living in other regions and countries with different Internet access, or a comparison of online buyers with respect to their individual characteristics (such as gender, age, cultural characteristics, available disposable income for tourism activities, accommodation preference, level of education etc) may produce different outcomes. Still, the present study may provide an interesting benchmark for the formulation of e-consumer behaviour and purchasing process of tourism and hospitality products online.

Highlights

- ☐ The chapter examines the attributes' complexity on tourism online purchasing intentions.
- ☐ It compares linearity with asymmetry using Qualitative Comparative Analysis.
- ☐ The research progresses from fit to predictive validity.
- ☐ Marketing, benefits/risks, price/quality, and trust produced complex statements.
- ☐ The results suggest that QCA is more suitable than regression in tourism research.

Questions

1. Why tourism and hospitality research is advisable to focus on complexity examination?
2. Does the examination of complexity in tourism and hospitality research make the linear analysis obsolete?
3. Can you identify a couple of aspects dealing with destination management that should be studied under the complexity perspective?

References

Aaker, D., & Day, G. (1990), *Marketing Research*. New York: Wiley.

Ahn, T.D., Ryu, S., & Han, I. (2004), The impact of the online and offline features on the user acceptance of Internet shopping malls, *Electronic Commerce Research and Applications*, 3, 405-20.

Allison, P.D. (2001), *Missing Data*. Thousand Oaks: Sage Publications.

Amaro, S., & Durate. P. (2015), An integrative model of consumers' intentions to purchase travel online, *Tourism Management*, 46, 64-79.

Armstrong, J. (2012), Illusions in regression analysis, *International Journal of Forecasting*, 28:3, 689-94.

Baggio, R. (2008), Symptoms of complexity in a tourism system, *Tourism Analysis*, 13(1), 1-20.

Bente, G., Baptist, O., & Leuschner, H. (2012), To buy or not to buy: Influence of seller photos and reputation on buyer trust and purchase behaviour, *International Journal of Human-Computer Studies*, 70(1), 1-13.

Bhatnagar, N., Lurie, N., & Zeithaml, V. (2003), Reasoning about online and offline service experiences: The role of domain-specificity in the formation of service expectations, *Advances in Consumer Research*, 30, 383-84.

Bjork, P., & Kauppinen-Raisanen, H. (2012), A netnographic examination of travelers' online discussions of risks, *Tourism Management Perspectives*, 1(2-3), 65-71.

Bore, I., Rutherford, C., Glasgow, S., Taheri, B., & Anthony, J. (2017), A systematic literature review on eWOM in the hotel industry: Current trends and suggestions for future research, *Hospitality and Society*, 7(1), 63-85.

Boukas, N., & Ziakas, V. (2014), A chaos theory perspective of destination crisis and sustainable tourism development in islands: the case of Cyprus, *Tourism Planning and Development*, 11(2), 191-208.

Bruce, N., Haruvy, E., & Rao, R. (2004), Seller rating, price, and default in online auctions, *Journal of Interactive Marketing*, 18(4), 37-50.

Chai, A. (2012), Consumer specialization and the demand for novelty: A reconsideration of the links and implications for studying fashion cycles in tourism, *Jahrbücher für Nationalökonomie und Statistik*, 232(6), 678 701.

Chikweche, T. & Fletcher, R. (2010), Understanding factors that influence purchases in subsistence markets, *Journal of Business Research*, 63(6), 643-50.

Dholakia, U.M. (2001), A motivational process model of product involvement and consumer risk perception, *European Journal of Marketing*, 35(11-12), 1340-60.

Dibb, S., & Simkin, L. (2013), *Marketing Essentials*. Cengage Learning.

Dylan, S. (2017), On shipwrecks and sea nymphs: Fragments of Maltese hospitality, *Hospitality and Society*, 7(1), .3-18.

Evans, J.R., & Mathur, A. (2005), The value of online surveys, *Internet Research*, 15(2), 195-219.

Faulkner, B. & Russell, R. (2000), Turbulence, chaos and complexity in tourism systems: A research direction for the new millennium, In R. Faulkner, G. Moscardo, & E. Laws (Ed.), *Tourism in the 21st Century: Lessons from experience* (pp.328-49). London: Continuum.

Faulkner, B., & Russell, R. (1997), Chaos and complexity in tourism: In search of a new perspective, *Pacific Tourism Review,* 1, 93-102.

Garbarino, E., and Maxwell, S. (2010), Consumer response to norm-breaking pricing events in e-commerce, *Journal of Business Research,* 63(9-10), 1066-72.

Gefen, D., Karahanna, E., & Straub, D.W. (2003), Trust and TAM in online shopping: An integrated model, *MIS Quarterly,* 27(1), 51-90.

Gigerenzer, G., & Brighton, H. (2009), Homo heuristics: Why biased minds make better inferences, *Topics in Cognitive Science,* 1, 107-43.

Gross, M.J., & Brown, G. (2008), An empirical structural model of tourists and places: progressing involvement and place attachment into tourism, *Tourism Management,* 29(6), 1141-51.

Hong, I.B., & Cha, H.S. (2013), The mediating role of consumer trust in an online merchant in predicting purchase intention, *International Journal of Information Management,* 33(6), 927-39.

Hong, I.B., & Cho, H. (2011), The impact of consumer trust on attitudinal loyalty and purchase intentions in B2C e-marketplaces: Intermediary trust vs seller trust, *International Journal of Information Management,* 31(5), 469-79.

Hong, Z., & Yi, L. (2012), Research on the influence of perceived risk in consumer on-line purchasing decision, *Physics Procedia,* 24(B), 1304-10.

Hyun, M. (2006), *The effects of tourism information web site factors on usefulness, web site attitude and behavior*. PhD Thesis, Seoul: Sejong University.

Jansen, B.J., Ciamacca, C.C., & Spink, A. (2008), An analysis of travel information searching on the web, *Journal of Information Technology and Tourism,* 10(2), 101-18.

Keating, B.W., Quazi, A.M., & Kriz, A. (2009), Financial risk and its impact on new purchasing behavior in the online retail setting, *Electronic Markets,* 19(4), 237-50.

Kellert, S. (1993), *In the Wake of Chaos: Unpredictable order in dynamical systems*. Chicago IL: University of Chicago Press.

Kent, R.A., & Argouslidis, P.C. (2005), Shaping business decisions using fuzzy-set analysis, *Journal of Marketing Management,* 21(5-6), 641-58.

Kholoud Al-Qeisi, K., Dennis, C., Alamanos, E., & Jayawardhena, C. (2014), Website design quality and usage behavior: Unified theory of acceptance and use of technology, *Journal of Business Research,* 67(11), 2282-90.

Kim, I. (2007), The role of perceived risk on purchase intention in Internet shopping, *International Journal of Electronic Marketing and Retailing,* 1(3), 279-88.

Kim, D.J., Donald, L.F., & Rao, H.R. (2008), A trust-based consumer decision-making model in electronic commerce: the role of trust, perceived risk and their antecedents, *Decision Support Systems*, 44(2), 544-64.

Komiak, S.Y.X., & Benbasat, I. (2006), Understanding customer trust in agent-mediated electronic commerce, web-mediated electronic commerce, and traditional commerce, *Information Technology and Management*, 5(1-2), 181-207.

Kyle, G., Graefe, A., Manning, R., & Bacon, J. (2003), An examination of the relationships between leisure activity involvement and place attachment among hikers along the Appalachian Trail, *Journal of Leisure Research*, 35(3), 249-73.

Law, R., Leung, R., & Buhalis, D. (2009), Information technology applications in hospitality and tourism: a review of publications from 2005 to 2007, *Journal of Travel and Tourism Marketing*, 26(5-6), 599-623.

Laws, E. & Prideaux, B. (2005), Crisis management: A suggested typology, *Journal of Travel and Tourism Marketing*, 19(2-3), 1-8.

Leenders, M.A.A.M., & Wierenga, B. (2008), The effect of the marketing–RandD interface on new product performance: The critical role of resources and scope, *International Journal of Research in Marketing*. 25(1), 56-68.

Liang, T.P., & Jin-Shiang, H. (1998), An empirical study on consumer acceptance of products in electronic markets: a transaction cost model, *Decision Support Systems*, 24(1), 29-43.

Lichtenstein, D.R., Ridgway, N.M., & Netemeyer, R.G. (1993), Price perceptions and consumer shopping behavior: a field study, *Journal of Marketing Research*, 30(2), 234-45.

Longest, K., and Vaisey, S. (2008), Fuzzy: A program for performing Qualitative Comparative Analyses (QCA) in STATA, *The STATA Journal*, 8(1), 79-104.

Marom, O., and Seidmann, A. (2011), Using 'last-minute' sales for vertical differentiation on the Internet, *Decision Support Systems*, 51(4), .894-903.

McDonald, J.R. (2009), Complexity science: an alternative world view for understanding sustainable tourism development, *Journal of Sustainable Tourism*, 17(4), 455-71.

Montoya-Weiss, M., Voss, G., & Grewal, D. (2003), Determinants of online channel use and overall satisfaction with a relational, multichannel service provider, *Journal of the Academy of Marketing Science*, 31(4), 448-58.

Moran, G., & Muzellec, L. (2017), eWOM credibility on social networking sites: A framework, *Journal of Marketing Communications*, 23(2), 149-61.

Nerkar, A., & Roberts, P.W. (2004), Technological and product–market experience and the success of new product introductions in the pharmaceutical industry, *Strategic Management Journal*, 25(8-9), 779-99.

Nilson, T.H. (1995), *Chaos Marketing: How to win in a turbulent world.* London: McGraw-Hill.

No, E., & Kim, J.K. (2015), Comparing the attributes of online tourism information sources, *Computers in Human Behavior*, 50, 564-75.

Okazaki, S. (2008), Determinant factors of mobile-based word-of-mouth campaign referral among Japanese adolescents, *Psychology and Marketing*, 25(8), 714-31.

Olmedo, E., & Mateos, R. (2015), Quantitative characterization of chaordic tourist destination, *Tourism Management*, 47, 115-26.

Olya, H. G., Gazi, Z. A., Aksal, F. A., & Altinay, M. (2018). Behavioral intentions of disabled tourists for the use of peer-to-peer accommodations: An application of fsQCA. *International Journal of Contemporary Hospitality Management*, 30(1), 436-454.

Ordanini, A., Parasuraman, A., & Rubera, G. (2014), When the recipe is more important than the ingredients: A Qualitative Comparative Analysis (QCA) of service innovation configurations, *Journal of Service Research*, 17(2), 134-49.

Pan, B., & Fesenmaier, D.R. (2000), A typology of tourism related web sites: Its theoretical background and implications, *Information Technology and Tourism*, 3(3-4), 155-66.

Pantano, E., Iazzolino, G., & Migliano, G. (2013), Obsolescence risk in advanced technologies for retailing: A management perspective, *Journal of Retailing and Consumer Services*, 20(2), 225-33.

Papatheodorou, A. & Pappas, N. (2017), Economic recession, job vulnerability, and tourism decision making: a qualitative comparative analysis', *Journal of Travel Research*, 56(5), 663-77.

Pappas, N. (2014), The effect of distance, expenditure and culture on the expression of social status through tourism, *Tourism Planning and Development*, 11(4), 387-404.

Pappas, N. (2015), Marketing hospitality industry in an era of crisis, *Tourism Planning and Development*. 11(3), 333-49.

Pappas, N. (2017), The effect of marketing activities, benefits, risks, confusion due to over-choice, price, quality and consumer trust on online tourism purchasing, *Journal of Marketing Communications*, 23(2), 195-218.

Park, J., Gunn, F., & Han, S. (2012), Multidimensional trust building in e-retailing: Cross-cultural differences in trust formation and implications for perceived risk, *Journal of Retailing and Consumer Services*, 19(3), 304-12.

Pescher, C., Reichhart, P., & Spann, M. (2014), Consumer decision-making processes in mobile viral marketing campaigns, *Journal of Interactive Marketing*, 28(1), 43-54.

Ragin, C.C. (2000), *Fuzzy-set Social Science.* Chicago, IL: University of Chicago Press.

Roberts, S., & Pashler, H. (2000), How persuasive is a good fit? A comment on theory testing, *Psychological Review*, 107, 358-67.

Russell, R. & Faulkner, B. (2004), Entrepreneurship, chaos and the tourism area lifecycle, *Annals of Tourism Research*, 31(3), 556-79.

Sekaran, U., & Bougie, R. (2009), *Research Methods for Business: A skill-building approach.* Chichester: John Wiley.

Sevgin, A., Peristianis, N., & Warner, J. (1996), Residents' attitudes to tourism development: the case of Cyprus, *Tourism Management*, 17(7), 481-94.

Sigala, M. (2004), Designing experiential websites in tourism and hospitality: A customer-centric value approach, *International conference in information and communication technologies in tourism 2004*, International Federation of Information Technologies in Tourism (IFITT), 27-29 January, Cairo, Egypt.

Skarmeas, D., Leonidou, C.N., & Saridakis, C. (2014), Examining the role of CSR skepticism using fuzzy-set qualitative comparative analysis, *Journal of Business Research*, 67, 1796-805.

Statista (2016), *Worldwide online travel sales volume from 2010 to 2016.* http://www.statista.com/statistics/247321/forecast-of-worldwide-online-travel-sales-volume/ (accessed 13/04/2018)

Tarnanidis, T., Owusu-Frimpong, N., Nwankwo, S., & Omar, M. (2015), A confirmatory factor analysis of consumer styles inventory: Evidence from Greece, *Journal of Retailing and Consumer Services*, 22, 164-77.

Taylor, D., & Strutton, D. (2010), Has e-marketing come of age? Modelling historical influences on post-adoption era Internet consumer behaviours, *Journal of Business Research*, 63(9-10), 950-56.

Van der Meer, R.B., J. Quigley, & J.E. Storbeck. (2005), Using regression analysis to model the performance of UK coastguard centres, *Journal of the Operational Research Society*, 56(6), 630-41.

Woodside, A.G. (2013), Moving beyond multiple regression analysis to algorithms: Calling for adoption of a paradigm shift from symmetric to asymmetric thinking in data analysis and crafting theory, *Journal of Business Research*, 66(4), 463-72.

Woodside, A.G. (2014), Embrace•perform•model: Complexity theory, contrarian case analysis, and multiple realities, *Journal of Business Research*, 67, 2495-503.

Woodside, A.G., & Zhang, M. (2013), Cultural diversity and marketing transactions: Are market integration, large community size, and world religions necessary for fairness in ephemeral exchanges?, *Psychology and Marketing*, 30(3), 263-76.

Wu, P.L., Yeh, S.S., Huan, T.C., & Woodside, A.G. (2014), Applying complexity theory to deepen service dominant logic: Configural analysis of customer experience-and-outcome assessments of professional services for personal transformations, *Journal of Business Research,* 67(8), 1647-70.

Xiang, Z., Magnini, V.P., & Fesenmaier, D.R. (2015), Information technology and consumer behavior in travel and tourism: Insights from travel planning using the internet, *Journal of Retailing and Consumer Services*, 22, 244-49.

Xiang, Z., Wöber, K., & Fesenmaier, D.R. (2008), Representation of the online tourism domain in search engines, *Journal of Travel Research,* 47(2), 137-50.

Yang, T.C., & Lai. H. (2006), Comparison of product bundling strategies on different online shopping behaviors, *Electronic Commerce Research and Applications,* 5(4), 295-304.

Zahra, A., & Ryan, C. (2007), From chaos to cohesion - Complexity in tourism structures: An analysis of New Zealand's regional tourism organizations, *Tourism Management,* 28(3), 854-62.

Zeithaml, V.A. (1988), Consumer perceptions of price, quality, and value: A means – end model and synthesis of evidence, *Journal of Marketing,* 52(3), 2-22.

Zhang, K.Z.K., Cheung, C.N.K., & Lee, M.K.O. (2014), Examining the moderating effect of inconsistent reviews and its gender differences on consumers' online shopping decision, *International Journal of Information Management,* 34(2), 89-98.

12 Creating a better future for the hospitality and tourism education

Christina Dimitriou

Abstract

In recent years, countless hospitality and tourism programs have been experiencing low enrollment, budgetary issues, low scores in the college rankings, and failure to secure good job opportunities for their graduates. This occurs due to the implementation of unethical hiring and recruiting practices for faculty and staff, poor decision making over key operational areas, abusive supervision, lack of ethical leadership, and careless strategic planning. However, the major reason is that there has been a huge gap between the hospitality and tourism academia and the industry which keeps growing. This is a rapidly changing industry where new trends are constantly emerging, new technologies are introduced, and innovative strategies arise. Meanwhile, hospitality and tourism programs around the world fail to understand the ever-changing industry trends, adapt to the new standards, and revise their curriculum accordingly. COVID-19 was the final blow that pushed academic institutions to their limits and created additional challenges. This chapter aims to address the issues that the hospitality and tourism academia has been struggling with, offer practical recommendations that will help tackle its operational difficulties, provide top quality education, prepare the leaders of tomorrow effectively, and bring it closer to the industry.

Keywords:

Hospitality and tourism education, hospitality and tourism industry trends, hospitality and tourism curriculum design, unethical leadership, equality, diversity

Introduction

One of the major obstacles that has been standing in the way of success of the hospitality and tourism academia throughout the years is the huge gap that separates it from the industry. There is a clear disconnect between the industry's needs and how academia prepares its hospitality and tourism students for a successful career upon graduation. That has a direct impact on the quality of the curriculum and the education that our hospitality and tourism programs offer which, in turn, influences their enrollment and financial performance. There are many reasons for this which include unethical hiring and recruiting practices for faculty and staff, poor decision-making over key operational areas, abusive supervision, and lack of ethical leadership. While our hospitality and tourism programs have been struggling to deal with these serious problems, the COVID-19 pandemic, brought a series of new challenges that most of them were not prepared or ready to face. Now, more than ever, is the time to learn from the mistakes of the past and help the academic world of hospitality and tourism make a brand-new start that will lead to success. The purpose of this chapter is to highlight those areas that need improving and offer recommendations on what can be done to improve our hospitality and tourism programs, help them achieve their goals, and create a better future for our field. It is our responsibility as educators to prepare ethical future leaders and course designers must collaborate with those working in hospitality to keep up with the industry's needs and prepare their graduates accordingly (Dimitriou, 2017).

A review of hospitality and tourism education

Many hospitality and tourism programs are facing a number of critical challenges such as budgetary and low enrollment issues, high faculty turnover, and low scores in the college rankings. This indicates that we are on the wrong path and are engaging in poor practices and procedures that hinder success. There are certain areas in our hospitality and tourism programs around the world that need to be improved, if we want to achieve excellence and offer high quality education to our students. It is important to identify and address these areas which negatively impact our schools, departments and programs. These areas are presented in Table 12.1.

Table 12.1: Areas for improvement of the hospitality and tourism education

Common mistakes in the recruiting and hiring process
The gap between the hospitality and tourism industry and academia
Ignoring the hottest industry trends
Unethical leadership and abusive supervision
COVID-19 and its challenges

Common mistakes in the recruiting and hiring process

Sadly, a growing number of universities engage in nepotism and favoritism. These are "*unprofessional practices giving preferential treatment to relatives and friends in employment*" (Arasli & Tumer, 2008: 1237) who may not have the competencies needed to get that job. Hornby et al. (1974) define nepotism as "*the giving of special favour (esp employment) by a person in high position to his relatives*" (p. 566). According to Arasli and Tumer (2008), "*favoritism means the provision of special privilege to friends, colleagues and acquaintances, in the areas of employment, career and personnel decisions*" and both nepotism and favoritism "*easily penetrate and corrupt decisions and practices*" (p. 1239). They also supported that "*nepotism has negative effects on human resource management recruitment and selection practices because candidates are chosen for their friendships and blood relationships. Since those workers may not have adequate relevant knowledge and job skills, they may not perform as well as suitably qualified candidates. This destroys organizational fairness, motivation and harmony while bringing inefficiency along with insufficient training and development activities. Conflict may arise between the worker who is already in the job and the new member of staff who has been hired because of his/her family relationship*" (Arasli & Tumer, 2008: 1239). It is extremely hard to create a work environment where professionalism and institutionalization prevail when the workforce consists of relatives, friends, and people who share strong personal relationships. Things become more complicated when these people who are related (e.g., a married couple) work for the same department of a university and one of the spouses acquires a position of authority (e.g., department chair) in that department and becomes their immediate supervisor. In this scenario, it is easy to understand how this unethical climate can impact the department resulting in reduced morale, motivation, and productiveness of the other faculty members.

Another serious issue in hospitality and tourism education is hiring people who have nothing to do with the field; do not hold a single degree in this field, or with degrees totally unrelated to the world of management (e.g., arts and musical performance, geography, etc.). Moreover, having never worked in our industry, do not have industry contacts or links in our field. Yet, they are hired to teach, run our programs and departments, and join committees where they are expected to make important decisions about the course of our institutions and the hospitality and tourism education of our students. Another obstacle that stands in the way of effective hospitality and tourism education is hiring people that lack passion, excitement, or genuine love for teaching. They join academia for all the wrong reasons (e.g., long vacation, paid travel for conferences, etc.). How can we effectively prepare the hospitality and tourism leaders of tomorrow by equipping our programs with faculty who lack knowledge, experience, background, and passion about our field?

The gap between the industry and academia

Hiring the wrong people not only creates serious operational problems inside the institution, but also plays a key role in increasing the gap between industry and academia. First and foremost, these people alienate hospitality and management students from the industry. Since they are not familiar with the industry needs and trends, they are unable to teach students how the industry works and what its needs are. They also fail to provide students with the right tools on how to build a successful career upon graduation and how to deal with the challenges and ethical dilemmas they may face. Their lack of a good industry network and key contacts that could guide, mentor, or provide potential jobs to their students puts the hospitality and tourism program at a disadvantage and creates further problems in the long-run. The industry got so disappointed with the poor and inadequate preparation and poor performance of our students that many hotel chains developed their own training programs to educate and prepare their management and staff for success.

Ignoring the hottest industry trends

Any expert in the field knows that the hospitality and tourism industry is truly international, extremely diverse, and constantly changing and evolving. Understanding the industry and keeping up with its trends is a huge factor that contributes to the success of a hospitality and tourism program. Failure to keep up with current trends leads to poor curriculum decisions and design, outdated taught courses, and directly impacts the enrollment rates which, in turn, lead to financial problems and budgetary issues.

The hottest trend for the tourism industry is managing successfully the increasing number of moral issues and dilemmas it is facing. In recent years, this very large industry has been struggling to support moral standards within its ranks. The United Nations World Tourism Organization's Global Code of Ethics for Tourism along with other codes of ethics provided by various tourism segments served as its major sources (Dimitriou, 2017). "*However, it is quite questionable to what extent these sources are adequate to cover critical issues and deal with morals which vary so significantly from culture to culture and from destination to destination*" (Dimitriou, 2017: 46). On top of that, "*the tourism industry is still trying to figure out how to effectively implement sustainable tourism and those principles emphasizing moral issues*" (Dimitriou, 2017: 45). Our industry needs, now more than ever, moral and responsible tourist leaders who will promote ethical and sound decision-making, and foster diversity, equity, and inclusion in our tourist businesses, local communities, and governments. The same applies to the hotel industry which is involved in an increasing number of ethical scandals that lead to expensive lawsuits and other detrimental consequences. Examples include the pregnancy and disability lawsuit of the Verona Resort and Spa where a female front desk employee with pregnancy-related and diabetic conditions got fired instead of having her needs accommodated (U.S. Equal Employment Opportunity Commission, 2019a) and the pay discrimination lawsuit against the Courtyard Monroe Airport hotel in Monroe, LA where women were paid less because of their gender (U.S. Equal Employment Opportunity Commission, 2019b).

Other important trends include the issues of overtourism, climate change, and global warming and their impacts on the tourism industry in the long-run. "*Overtourism is a new term for an old problem, namely, excessive numbers of tourists at a specific destination that can result in negative impacts of all types on the community involved*" (Dodds & Butler, 2019). Global warming "*refers to the effect on the climate of human activities, in particular the burning of fossil fuels (coal, oil and gas) and large-scale deforestation, which cause emissions to the atmosphere of large amounts of 'greenhouse gases', of which the most important is carbon dioxide" and is the most important environmental problem the world faces*" (Houghton, 2005: 1343).

The latest trends, though, are associated with the technological advances that the hospitality and tourism industry is trying to incorporate in its operations. These include robots, chatbots, and automation. HiltonWorldwide introduced Connie, a robot concierge, Starwood Hotels and Resorts uses a Botlr which delivers room service and the Yotel in New York has a luggage toting robot. (Trejos, 2016). Many travelers are using internet chatbots to book their travel and accommodation. These are carefully designed Artificial Intelligence (AI) systems that can answer questions and provide useful information to guests when human operators are unavailable (Revfine, 2021). As our customers are becoming more tech-savvy, there is an increasing demand for digitalization, smart use of technology, and personalized service. Our students need to know what is going on in the industry and how to adapt to the changes that are taking place. Therefore, it is the responsibility of our educators to stay current with the ever-changing industry trends, bring them in the classroom, and incorporate them in their courses and curriculum. That said, how many hospitality and tourism educators actually do so and how many of the offered hospitality and tourism programs focus on this?

Unethical leadership and abusive supervision

The major reason that many universities face serious financial problems, enrollment issues and low scores on academic rankings is the lack of ethical leadership and sound decision-making. Ethical leadership requires that we put the interests of the department and the university first. "*Ethical leaders do not compromise their ethical standards and principles under any circumstances. They are free of biases and prejudices and never put their self-interests first. They do not merely preach about ethics or view it as just another job responsibility. This is their way of life and the values they set for their companies are the very values they believe in and abide by in their own lives*" (Dimitriou & Schwepker, 2019: 671). Too often, though, hospitality and tourism administrators place emphasis on satisfying personal interests, basing their decisions on what will make them more popular, not what will benefit their programs and departments. Outstanding and fair leaders respect and value the role of teamwork and understand that if their team is weak, they are weak. They hire the best candidates for their departments. However, some academic leaders fail to recognize that, considering talented and qualified candidates as threats and not hiring them. This is a key characteristic of unethical leadership. Unethical leadership refers to "*behaviors conducted and decisions made by organizational leaders that are illegal and/or violate moral standards, and those that impose processes and structures that promote unethical conduct by followers*" (Brown & Mitchell, 2010, p. 11).

Another issue, in recent years, has been high faculty turnover due primarily to unethical leadership. This is consistent with a basic management principle that we cannot treat people in an awful manner and expect them to stick around. Many studies (DeConinck, 2014; Elçi et al., 2012; Shafique et al., 2018; Suifan et al., 2020) have found that ethical leadership negatively influences turnover intention. Bies and Tripp (1998) view abusive supervision as a source of injustice which causes significant problems to organizations and employees. According to Zellars et al. (2002), "*a small but growing body of empirical research suggests that abused subordinates report greater job and life dissatisfaction, intentions to quit their jobs, role conflict, and psychological distress, compared with their non-abused counterparts, and that subordinates' perceptions of unfairness explain their responses to abusive supervision*" (p. 1068). Dorenkamp and Weiß (2018) contended that "*a growing number of postdoctoral academics cite stressful working conditions for considering abandoning their studies and leaving the academic profession entirely before they obtain a tenured position*" (p. 747). Johnsrud and Rosser (2002) argued that "*searches are costly, and too often the faculty who leave are those the institution would prefer to retain*" (p. 518). So, apart from money, our hospitality and tourism programs are also losing all the talent, experience, expertise, high level of professionalism and commitment to excellence that these faculty have. They lose great opportunities to advance the quality of education they offer, improve their product, services, research, and image. How can we provide high quality education when we are unable to maintain educated, experienced, dedicated, top-quality faculty who excel in our field? How can we be talking about promoting diversity, equality and inclusion when we disrespect our faculty so bad?

COVID-19 and its challenges

Some hospitality and department programs had already incorporated a form of online learning in their curriculum, while the majority of them were only focused on face-to-face education. So, when COVID-19 hit and there was a sudden need to move everything online, many universities were completely unprepared for such a transition. Due to the fact that everything happened so fast, the majority of institutions "*shifted to remote teaching or distance learning in the simplest and most accessible ways possible — via basic tech tools on learning-management systems, web conferencing platforms, email, and phone*" (Bessette et al., 2020: 1). Those educators who were more knowledgeable and familiar with online teaching stepped up to help others through "*impromptu Facebook groups, public video tutorials, email lists, blog posts, collaboratively curated websites, and even offers of virtual class visits have popped up to offer support to faculty members at any type of institution*" (Bessette et al., 2020). The biggest challenges were not only related to teaching an existing course that was already in progress. Things got much more complicated with the design, and development of a new course. This is a procedure that takes somewhere between 6 months to a year to properly organize and prepare it. "*In that process, faculty members work with instructional designers, technologists, and others on content delivery, assignment design, and assessment*" (Bessette et al., 2020: 1). On numerous occasions, things got done on a much more superficial level jeopardizing the quality of the outcome. "*Just as all instructors are not successful in the classroom setting, not all will be successful online*" (Palloff & Pratt, 2007: 7). Expert online instructors will confirm how time-consuming and how rigorous it is to teach a course online. The educator has to be present, active, and involved at all times and offer synchronous and asynchronous sessions to teach this course. However, many faculty members found themselves working from home and barely posting any lectures and material for students while there was hardly any 'live' interaction.

Strategies that could lead to a better future

As we have seen, so far, the hospitality and tourism academia has various problems to deal with which are quite complicated as they involve many different areas. Trying to address them all at once would be inadvisable. It is more prudent to divide them and address each and every one of them individually. One thing that must be made clear, though, is that in order to successfully implement the recommended strategies listed below, teamwork and collegiality is essential.

Our hospitality and tourism faculty

The most important part of a hospitality and tourism program or department is the set of people who are in it. The faculty and staff who work together to achieve its goals. So, the first step is to ensure that we hire the right people who match our organizational culture, values, and mentality and will be a good fit for our department; educators who are passionate about teaching and love this field; educators with strong pedagogical knowledge and skills.

Any skilled, experienced, and knowledgeable educational leader will confirm that we cannot hire mediocrity and expect excellence. By hiring the wrong people, we set ourselves up for failure. We need talented, qualified people who have what it takes to help us succeed and prosper. Hiring procedures must be based on clear-cut, transparent, unbiased, and objective criteria that apply to all candidates without exception. There are still a lot of people in hospitality and tourism academia who truly care, are willing, able, and committed to doing their best to make a difference and set a new standard for the field, the profession, and future generations.

There's also considerable pressure to manage the limited budget that hospitality and tourism administrators have to work with. Like any other business, we have to carefully evaluate our choices. Hiring someone on a lower salary is cheaper and may be appear cost-effective at first glance, but could severely risk the quality of the work and performance he/she would bring. Administrators must be able to see the big picture and plan for the long term with a clear strategy in mind. They need to know when it is appropriate to create a tenure-track versus a non-tenure track position in order to cover their needs. Recently, a number of business schools, adapted a strategy of hiring non-tenured faculty, because it was saving them money, but it opened the door to a lot of additional challenges (Gao et al., 2010). Opening up a position that requires experience and expertise, especially when a new program is being developed, should not be filled by an All-But-Dissertation (ABD) candidate with zero working experience and no teaching background. If we are a little more mindful of our choices and the planning of our programs and avoid making the mistakes of the past, we will start heading in the right direction.

Keeping up with the industry trends

Keeping up with the industry trends is an integral part of our work as hospitality educators that we must pay careful attention to. For example, identifying and carefully addressing all the travel and transportation issues that derived from the spread of COVID-19 has been a major challenge for the industry. The virus has forced all modes of travel to change the way they were doing business and adapt to the new standards. Safety and hygiene issues have become their top priority (Revfine, 2021). Adapting to the new practices that the experts recommend, such as respecting social distancing, providing antibacterial wipes, requiring everyone to wear face masks (even double-masking), and enforcing these rules, are critical parts of the strategy that will lead to safe travel again. The virus brought a shift from international to local travel as the various restrictions discouraged people to travel abroad (Revfine, 2021). Now, the industry focuses on promoting local destinations and targeting locals. That requires a different marketing approach and a new perspective in the way they do business. We need to use all the information and feedback we get from the industry wisely, and incorporate it in our classes to make key decisions on revising our courses and including it in our research.

Since the hospitality and tourism academia and industry rely so heavily on each other, establishing good relationships with hospitality and tourism organizations in our locality is a great way to start. Then, we must broaden our network through national and international associations and cultivate, nourish, and retain these relationships. We must connect our students with industry professionals and involve them in our classes and university life. Invite them as guest speakers, either virtually or in-person (when the situation allows it), and seek their opinion when we are designing a course in their field; connect them with graduating students who are looking for a job in their field and encourage them to mentor our students. Hospitality and tourism administrators must focus on having frequent dialogues with the industry to understand their needs; then share the valuable information they receive with the faculty, staff and students, and use it accordingly in order to improve the quality of the education they offer.

Our hospitality and tourism curriculum

The key to the success of a hospitality and tourism department is the product it offers. Therefore, the priority for hospitality and tourism administrators should be to design and develop a cutting-edge curriculum. In order to achieve that, taught courses should be revised and new ones introduced that reflect the industry's new standards as well as new trends that are emerging. For example, knowing that the industry is currently switching from international to local travel should be incorporated in tourism marketing courses and teach students the best strategies to accomplish that. Courses related to airlines, cruises, hotels, restaurants or bars should address the need for setting safety and hygiene standards due to COVID-19. They should teach students how to effectively plan and organize trips that are based on strict cleaning policies, socially distanced seating, providing anti-bacterial wipes and enforcing face masks and ensuring that passengers will comply with the new regulations.

According to Dimitriou (2020), *"a strong hospitality and tourism curriculum places emphasis on helping students gain a global perspective and learn how to think globally and act locally"* (p. 3) and is multicultural. Shiel (2007) feels that the focus should be on preparing our students to become global citizens. Hunter et al. (2006) argue that students should have global competencies.

Dimitriou (2020) recommends that a hospitality and tourism curriculum should concentrate on business ethics education as she supports that it is the educators' duty to teach students how to be ethical and contribute to a better future for this field. *"International hospitality business ethics should be introduced as a core course and taught with a hands-on approach that students from all different backgrounds can relate to and fully understand."* (Dimitriou, 2020, p. 1). Hospitality and tourism educators must promote an ethical climate in their classrooms where honesty, respect, open communication, and team spirit will prevail. We must also ensure that our students feel welcome and safe, and any inappropriate, offensive, and disruptive behavior is not tolerated. Our teaching should be enriched with a variety of different perspectives, so that our students are exposed to new ways of looking at the hospitality and tourism discipline and broaden their horizons.

The hospitality and tourism industry is the most diverse in the world and learning how to support and promote diversity starts in our classrooms. In fact, this is one of the best ways to prepare our hospitality and tourism students for a successful career in this industry which applies to both face-to-face and online education. Active participation strategies such as large group, small group, and individual activities should be implemented and rewarded in hospitality and tourism classes to increase student engagement. Online education offers various tools such as threaded discussions, video-conferencing, posting of blogs and videos that allow students to connect with one another, and help build a collegial and supportive online community where communication matters and strong bonds are cultivated. Case-based and cooperative learning activities should also be used. Furthermore, this time around, those who want to introduce a new classroom technology to their faculty must take the time to do it right. They need to start by creating awareness and providing adequate knowledge such as "*training workshops, internal marketing newsletters documenting and explaining best technology – based educational practices and even new educational technology trial opportunities to ensure that faculty members are knowledgeable about the new technology*" (Goh & Sigala, 2020: 162). There should also be adequate technical support available for those students, faculty, and staff who encounter difficulties using online tools and programs. Technical support gives people the confidence and sense of security they need to use the technology (Eteokleous, 2008).

Ethical leadership

Hospitality and tourism administrators must implement ethical leadership in their programs and departments where faculty, staff and students will feel free to express their opinions, thoughts, and ideas and come together as a team to work towards achieving the department and university goals. They must lead by example, support and promote innovation and mutual respect, open the door to creativity, and become exceptional role models to their subordinates as well as younger generations. They must "*walk the talk and talk the walk, patterning their behavior and organizational processes to meet moral standards*" (Brown & Mitchell, 2010: 584). It is imperative that transparent, fair, and just recruiting and hiring procedures are followed. Emphasis should also be placed on faculty retention by treating people with respect, valuing and celebrating their accomplishments, guiding them and supporting them to become the best they can be, and not allowing any room for unethical behavior.

The ultimate goal is to create and foster a healthy, harmonious, and ethical climate in our hospitality and tourism departments. The key to achieving that is by promoting ethics, diversity, inclusion, and treating everyone equally. Ahmad (2020) argued that: "*Equity in 2020 requires more than a diversity statement*" (p. 1). She contended that administrators must demonstrate a real commitment to equity and diversity, by stopping talking, doing some careful listening, taking detailed notes and confirming on what they have heard to avoid any misunderstandings and then put together proposals with specific action steps to be followed. The lines of communication should be open so that everyone involved in the process, is fully aware of the guidelines that must be followed and do their part to contribute to these efforts. "*Diversity and equity officers may serve in moderator or support roles, but these meetings must be between student groups and senior leadership. For this to work, the top administration must commit to direct dialogue and concrete action, and be open to tough challenges on funding, hiring decisions, access to services, and police presence on the campus*" (Ahmad, 2020: 1). Lawson-Borders and Perlmutter (2020) pointed out that "*plans for increasing faculty diversity must be feasible and practical as well as idealistic and ethical*" (p. 1). Their most effective strategies to increase diversity and inclusion in faculty-hiring practices included: identifying prospects and building the University's inclusive brand ahead of time, knowing what's appealing about the institution, program or place, encouraging hiring committees to recruit – not just 'open and advertise', widening the existing assumptions about the meaning of 'qualified', enlisting allies to promote the faculty search, rethinking the role of students in searches, redefining the notion of 'fit', and laying out the path to long-term success and not just the start-up package. Fortunately, several universities have already started placing emphasis on inclusion, diversity, and equality issues across their campuses and becoming more sensitive towards them.

Conclusion

A number of factors led to key operational problems that hospitality and tourism programs have been facing in recent years. The sudden COVID-19 outbreak exacerbated the situation, and brought chaos and confusion to the hospitality and tourism academia that has been much harder to control. The purpose of this chapter is to examine all these factors and recommend some strategies and solutions that could help the hospitality and tourism administrators deal with them successfully and get their institutions back on track.

One of the key take-aways is that ethical leadership is extremely important in the organizational behavior field and the hospitality and tourism academia. So, our administrators must establish an ethical framework in their programs and departments. In addition, they should include communication and interaction in their managerial values to make sure that everyone is on the same page, lessen any negative perceptions that those involved may have and actively convey them through their programs. Hiring practices should be re-evaluated and built on a more ethical and fair foundation. Honest, clear, and transparent procedures must be put in place that support and encourage ethical behavior and broad-mindedness. Bringing knowledgeable and experienced people on board will help us build a cutting-edge curriculum and create a better product for our students.

Efforts must be made to connect with the industry and cultivate closer relationships that will benefit both parties. Emphasis must be placed on staying current with industry trends and adapting to new standards. A successful hospitality and tourism curriculum is aligned with current industry trends, offers students strong industry knowledge and experience, teaches them excellent leadership and communication skills, places emphasis on business ethics and professionalism, and reflects the real-life situation that they will experience upon graduation (Dimitriou, 2020). This will attract talented, ambitious, motivated, and top-quality students. Enrollment rates will increase. Students will be graduating on time with high grades and will become the ambassadors of our university. Industry professionals will realize that we understand their needs and are capable of providing them with quality graduates. Our programs will score much higher on academic rankings. Our hospitality and tourism education needs fair and ethical administrators who are broad-minded, free of biases and prejudices and promote diversity, inclusion and justice in their departments. A healthy ethical and supportive work environment will reduce faculty turnover.

Adapting smoothly and effectively to online education and taking all the right steps to make it right is vital. We already have some wonderful hospitality and tourism faculty and administrators who are both willing and committed to do everything they can to overcome these challenges. Some universities have already started taking action and moving towards the path of equality, diversity, inclusion, and ethical leadership. This will have a tremendous impact on the academic institution's image and reputation. It will also signal to the industry that we do respect these values; we know exactly how to foster an ethical climate in our campuses and create ethical and responsible leaders. It is only a matter of time before things start changing and open up a new chapter in our post COVID-19 era.

Highlights

- ☐ Efforts must be made to connect with the industry and cultivate closer relationships that will benefit both parties and bridge this existing gap.
- ☐ Ethical leadership is the key that will help the hospitality and tourism academia overcome its operational problems and set the foundation for a more prosperous, successful, and promising future.
- ☐ Hospitality and tourism administrators must create and foster a healthy, harmonious, and ethical climate in their departments. The key to achieving that is by promoting ethics, diversity, inclusion, and treating everyone equally.
- ☐ Recruiting and hiring procedures must be based on clear-cut, transparent, unbiased, and objective criteria that will apply to every candidate without any exceptions.
- ☐ The priority of hospitality and tourism administrators should be to design and develop a cutting-edge curriculum which is aligned with current industry trends, offers students strong industry knowledge and experience, teaches them excellent leadership and communication skills, places emphasis on business ethics and professionalism, and reflects the real-life situation that they will experience upon graduation.

Questions

1. What are some online teaching tools and strategies that work best for you? Why?
2. If you were assigned to lead a hospitality and tourism management department, what policies and practices would you use in order to be an ethical leader?
3. What are some effective ways that you would use to keep up with the current industry trends? How would you use the information you have found to advance your career?

Case Study: The unethical search committee chair

You are a search committee member for a full-time tenure-track faculty position that has opened in your hospitality and tourism management department. You, along with the other three committee members and the Search Committee Chair created the rubric which includes the criteria that you will use to evaluate each and every one of the applicants. You received a total of twenty-two applications. You have had two meetings, so far. During your first meeting, you decided which eight candidates you will interview for the position. On your second meeting, you narrowed them down to three. Now, you are having your final meeting to decide which one is the most appropriate for the position. Based on the results of your graded rubrics, you and the other three committee members select one specific candidate. Clearly, he is the best choice as his knowledge and experience fulfill all the requirements included in the job posting. Plus, he is very passionate about teaching and seems very excited to join the department. However, when you turn to the Search Committee Chair to get his opinion, he is not happy with this decision. He says to all of you: "Yes, he is a strong candidate for this position but, for reasons better left unsaid, I want to bring another candidate". The person he picks has much less teaching experience and number of publications, no work experience in the hospitality and tourism industry, and no pedagogical education. You are not at all surprised, though, because when you did a little more 'digging' as you were examining the credentials and background of the three finalists, you discovered that the candidate that the Search Committee Chair is in favor of happens to be one of his classmates from graduate school and that they have been close friends for many years. Meanwhile, everyone in this meeting knows very well that at the beginning of the semester, it was made very clear to all faculty and staff that ethical and fair procedures would apply on all university operations and there would be zero-tolerance for any violators.

Questions:

1. What do you do?
2. Who do you involve?

References

Ahmad, A. S. (2020) Equity in 2020 requires more than a diversity statement, *The Chronicle of Higher Education*. https://community.chronicle.com/news/2360-equity-in-2020-requires-more-than-a-diversity-statement

Arasli, H. & Tumer, M. (2008) Nepotism, favoritism and cronyism: A study of their effects on job stress and job satisfaction in the banking industry of North Cyprus, *Social Behavior and Personality*, 36(9), 1237-1250.

Bessette, L. S., Chick, N. & Friberg, J. C. (2020) 5 myths about remote teaching in the Covid-19 crisis, *The Chronicle of Higher Education*. https://community.chronicle.com/news/2346-5-myths-about-remote-teaching-in-the-covid-19-crisis

Brown, M. E. & Mitchell, M. S. (2010) Ethical and unethical leadership: Exploring new avenues for future research, *Business Ethics Quarterly*, 20(4), 583-616.

Bies, R. J. & Tripp, T. M. (1998) Two faces of the powerless: Coping with tyranny, in R. M. Kramer & M. A. Neale (Eds.), *Power and Influence in Organizations*, Thousand Oaks, CA: Sage, pp. 203–219.

Brown, M. E. & Mitchell, T. R. (2010) Ethical and unethical leadership: Exploring new avenues for future research, *Business Ethics Quarterly*, 20(4), 583–616.

DeConinck, J. B. (2014) Outcomes of ethical leadership among salespeople, *Journal of Business Research*, 68(5), 1086–1093.

Dimitriou, C. K. (2017). The quest for a practical approach to morality and the tourism industry. *Journal of Hospitality and Tourism Management*, 31, 45-51.

Dimitriou, C. K. (2017). The quest for a practical approach to morality and the tourism industry. *Journal of Hospitality and Tourism Management*, 31, 45-51.

Dimitriou, C. K. (2020) Adjusting our teaching strategies, material, and techniques to the new standards, *The Tourism Education Futures Initiatives* (TEFI) 11-What's Tourism got to do with it? York St. John University, UK.

Dimitriou, C. K. & Schwepker, C. H., Jr. (2019) Enhancing the lodging experience through ethical leadership, *International Journal of Contemporary Hospitality Management*, 31(2), 669-690.

Dodds, R., & Butler, R. (2019) The phenomena of overtourism, *International Journal of Tourism Cities*, 5(4), 519-528.

Dorenkamp, I. & Weiß, E. E. (2018) What makes them leave? A path model of postdocs' intentions to leave academia, *Higher Education*, 75(5), 747-767.

Elçi, E,. Şener, I., Aksoy, S. & Alpkan, L. (2012) The impact of ethical leadership effectiveness on employees' turnover intention: The mediating role of work related stress, *Procedia-Social and Behavioral Sciences*, 58, 289-297.

Eteokleous, N. (2008) Evaluating computer technology integration in a centralized school system, *Computers & Education*, 51(2), 669–686.

Gao, T., Sirgy, M. J. & Johar, J. S. (2010) Developing a measure to capture marketing faculty's perceptions of unethical behavior, *Journal of Business Research*, 63(4), 366-371.

Goh, E. & Sigala, M. (2020) Integrating Information & Communication Technologies (ICT) into classroom instruction: teaching tips for hospitality educators from a diffusion of innovation approach, *Journal of Teaching in Travel & Tourism*, 20(2),156-165.

Hornsby, A. S., Cowie, A. P. & Gimson, A. C. (1974) *Oxford Advanced Learner's Dictionary of Current English*, Oxford, U.K.: Oxford University Press.

Houghton, J. (2005) Global Warming, *Reports on Progress in Physics*, 68, 1343-1403.

Hunter, B., White, G. & Godbey, J. (2006) What does it mean to be globally competent?, *Journal of Studies in Higher Education*, 10(3), 267-285.

Johnsrud, L. K. & Rosser, V. J. (2002) Faculty members' morale and their intention to leave: A multilevel explanation, *Journal of Higher Education*, 73(4), 518-542.

Lawson-Borders, G. & Perlmutter, D. D. (2020) 8 Practical, Sustainable Steps to a Diverse Faculty, *Chronicle of Higher Education*. https://community.chronicle.com/news/2412-8-practical-sustainable-steps-to-a-diverse-faculty

Palloff, R. M. & Pratt, K. (2007) *Building Online Learning Communities: Effective strategies for the virtual classroom, 2nd edn*, San Francisco, CA: John Wiley.

Revfine (2021) Tourism trends: The latest opportunities for the tourism industry. https://www.revfine.com/tourism-trends/

Shafique, I., Kalyar, M. & Ahmad, B. (2018) The nexus of ethical leadership, job performance, and turnover intention: The mediating role of job satisfaction, *Interdisciplinary Description of Complex Systems*, 16(1), 71–87.

Shiel, C. (2007) Developing global citizens: The way forward?, *International Journal of Learning*, 14(4), 153-167.

Suifan, T. S., Diab, H., Alhyari, S. & Sweis, R. J. S. (2020) Does ethical leadership reduce turnover intention? The mediating effects of psychological empowerment and organizational identification, *Journal of Human Behavior in the Social Environment*, 30(4), 410-428.

Trejos, N. (2016) Introducing Connie, Hilton's new robot concierge, USA TODAY. https://www.usatoday.com/story/travel/roadwarriorvoices/2016/03/09/introducing-connie-hiltons-new-robot-concierge/81525924/

U.S. Equal Employment Opportunity Commission. (2019a) Verona Resort and Spa to pay $16,000 to settle EEOC pregnancy and disability suit. https://www.eeoc.gov/eeoc/newsroom/release/5-13-19.cfm

U.S. Equal Employment Opportunity Commission. (2019b) EEOC sues Monroe Hotel for pay discrimination. https://www.eeoc.gov/eeoc/newsroom/release/7-22-19.cfm

Zellars, K. L., Tepper, B. T. & Duffy, M. K. (2002) Abusive supervision and subordinates' organizational citizenship behavior, *Journal of Applied Psychology*, 87(6), 1068–1076.

13

The perspectives of risk management in the Greek tourism industry: The case of the new services establishment

Maria Zoi Spanaki

Abstract

This study aims to highlight the meaning of the risk management sector and the importance of recognizing it as an essential field for the global economy. This study highlights the necessity of a new services risk management model to be established in the Greek tourism industry. Through an analysis of innovation and its impacts on economic growth, the study confirms the importance of accepting new services as a beneficial challenge which will create a new perspective for the tourism industry. Taking into consideration the limited literature sources about the topic, the contribution of this research is reflected on the recommended alteration of risk management model steps to new service management model. The literature gap is proposed to be covered through an analysis of innovation processes and their correlation with the importance of establishing new services and products into Greek tourism.

Keywords:

Risk management, innovation, Greek tourism, new services establishment

Introduction

In a continuously changing world, where technology shows rapid development and innovation is the key for global economic growth, the ability to identify, evaluate and assess risks is one of the most essential processes for all businesses in all sectors. Despite the first appearance of risk management in the early 1950s, the recognition of risk management as an important sector with its own department within businesses, was not established until the late 1990s with the creation of boards consisting of specialized risk management-oriented members. Risk management includes the process of identification, evaluation, monitoring of the risk and its assessment, and yields desirable results. Innovation is directly associated with a range of risks, as every new idea can be beneficial for a business but at the same time risky for potential losses until it is fully established.

Tourism is a huge industry, made up of individual sectors. For this reason, innovation is the basis for the improvement of the tourism industry by keeping a continuously competitive environment that makes tourism businesses put an constant effort to maintain a powerful presence within the arena. For success though, important components are the plan that a business follows for improvement and the strategies to deal with potential risks. The argument at the centre of this study is that in order for a sector to be successful, an individual risk management model should be created and followed step-by-step.

Below, information is presented about the connection of tourism with innovation through new services, focusing on Greece as one of the major tourism destinations globally. Despite that tourism is the strongest component for the Greek national GDP, there is still a need for strategies for development to deal with the establishment of new services. The recommendation for such a model is implemented in this study and the need for the creation of more efficient strategies is a clear step which will drive Greek tourism to success.

Background of risk management

The existence of the risk management sector dates back to some years after World War II. Despite the current theories which recognize risk management as a necessary established sector for the development of the corporate world, the reason for its further analysis was due to its association with the use of market insurance to protect businesses from accidental losses (Dionne, 2013). During the gradual improvement and development of several market sectors, people realized that activities within the market were costly and they had to deal with more risks than were covered by that insurance. The further analysis of the sector contributed to the publication of academic texts that contained the explanation of pure risk and mentioned other types of risks that potentially influence business growth (Crockford, 1982).

The concept of risk management started to acquire high significance not only for companies but also for individuals. Self-protection activities affected the probabilities of losses or costs before they even arose and this contributed to the discovery of prevention activities that nowadays are part of the risk management process (Dionne, 2013). During the 1960s, the risk prevention process started to become more important for businesses and soon further studies were conducted about the ways that risk investigation should be directed to avoid potential damages. The continued market growth and the creation of more activities, resulted in the classification of risks into main categories such as financial risk, operational risk, compliance and regulatory risk and subcategories including the risk of reputation, organizational risk and credit risk (Hopkin, 2018). The recognition of risk types was an important risk measurement tool for that period, even if there was still lack of studies for the development of the risk management process.

Despite the progress of risk management between 1950-1990, it was only in the late 1990s when it started to get officially recognized as an important sector. In that period, boards of directors and specialists in risk management acquired their own departments within businesses and their role included the process of monitoring, examination and analysis of potential risks that a business's next decisions may encounter (Scaillet, 2003). The significant role and the success of those specialists created the position of Chief Risk Officer and in the early 2000s, the risk management department emerged in most businesses with the responsibility to take the final decision for new business strategies (Scaillet, 2003).

Risk management definition and main types

Since the time risk management has been established as a recognized sector, a variety of studies have been conducted to define its meaning. Thinking about the definition of risk and the link between the first step of identification of risk and the last step of offering a solution to avoid it, risk management is the sector responsible for the identification, evaluation and prioritization of risks occurring within businesses and industries (Kunreuther, 2002). Activities that the risk management process includes are monitoring, minimizing and controlling the probabilities of risks, and establishing the best process to deal with any potential losses or damages (Hopkin, 2018). As mentioned by Demir (2019), risk is one of the fundamental elements that influences people's behaviour, and it expresses the challenge to deal with unknown results (Lemos, 2020). As risk management includes the term 'management', the aim of the sector is the process of managing the unpredictable future under the reality of a continuously growing competitive market that is characterized by constant achievements (Seaman, 2020).

As it is recommended by Aven (2017), there are three main categories of risk: non-business, business risk and financial risk. The first type of risk is a result of external environment impacts that affect the internal situation of the business. The business risk refers to the possibility of insufficient profits due to changes in consumers' behaviour and preferences, while the financial risk, which is the most common type, is related to business's financial damages and losses (Diebold et al., 2012). The three categories of risk include six main subcategories: economic risks, compliance, security, reputation, operational risk and competition risk (Kendrick, 2009). Analysing the subcategories, there are more groups that risks can be classified into, and as each type of risk occurs with a different impact on a business or industry environment, it is important and challenging to establish an individual process of dealing with risk assessment every time (Lemos, 2020).

The Risk Management Model (RMM)

Taking into consideration the steps that are followed for dealing with a potential risk, the risk management model (RMM) has been established and is followed by businesses to deal with any type of possible risk. The original model consisted of five steps as the basis for risk assessment and its solution. Identification, analysis, evaluation, threatening and reviewing of risk are the five components and their interrelation defines the RMM as the systematic process of responding to risks related to general concepts (Blumenthal et al., 2018; Soroka-Potrzebna, 2018). For a better understanding of the process, further analysis of risk management is important; this includes the risk assessment process and risk control.

The risk assessment step includes risk analysis and risk evaluation. The level of identification is part of risk analysis as well as the process of risk estimation. Risk evaluation contains the process of threat assessment as in this stage risk is recognised in order to find solutions and alternative ways of dealing with it. The last step of risk management is the process of risk control and risk reduction. At this point, risk is monitored to confirm that implemented solutions are ideal for its avoidance, and the step includes the final decision-making for risk reduction. The last stage is characterized as the risk review process and it has high importance for an efficient business result (Siegrist, 2021). In the worst possible scenario, if the chosen way to avoid risk does not eliminate the damage, it is necessary to extend the period of monitoring the treatment of risk and examine the alternative options until risk avoidance reaches the expected result (Krzemien et.al, 2013).

Due to technological achievements and the rapid growth of market competition, the risk management model has to be adjusted to each risk type for better outcomes. There are thousands of alternative models depending on the business, the sector and the general background of the market. For example, the risk management model that is followed in the IT and cyber security sector, differs from the one that is used in the health sector as Kopp et al. (2017) mention; and the model that is used for dealing with an operational type of risk is different from the one that is used to deal with an entrepreneurial risk type.

Risk management in the tourism industry

The risk management model plays a significant role for the future of all sectors, especially tourism, which is considered as one of the vital elements of the modern economic system and is highly dependent on risks, perceptions and crises. This is a complicated field to analyse, as risks that characterize the tourism sector are classified in different categories and affect the industry in many ways. Taking into consideration the fact that the tourism industry is a sum of individual sectors, such travelling, marketing, human resources and business management, it is logical to suggest that the outcome of tourism risks should be studied on an individual level and be dealt with as individual processes.

According to the World Tourism Organisation, there are four main sources of risks in the tourism industry (Štetic, 2017). First, there are the risks which are created by the tourism sectoritself and are related to commercial sources, which cover all non-political risks. These can be the compliance risk (regulation, organization and reputation of a business), the financial risk (economic situation within business that occurs from its organization) or the operational risk (process, employees, internal functions and operations). Another source of risk in the tourism industry is the human resources and institutional environment that includes training and competency of staff, injuries within working environments, conflicts, employee benefits liability; as well as risks from the external environment such as wars, social conflicts, terrorism, crime and human trafficking. The personal risk is also categorized as one of the main four sources and is related to the individual perception of tourists, who may be the reason for a conflict with local residents as they may cause a crime, disobey the local regulations and visit places where their safety may be at risk. The last source is the one which connects tourism with physical risks resulting from the environment. Climate changes, earthquakes, tsunamis or pandemics are classified in this source and their risk has become more intense each day in tourism destinations (da Rocha & da Silveira, 2021).

The success of a destination depends on being aware of tourist risks, as success is directly associated with the safety and security the place conveys to visitors (Zou & Zhu, 2020). The creation of a safe tourist environment can be a strong component for the destination to keep itself powerful among the global competition, as it is important for travellers to know in advance that they will be able to visit with no actual risk for their lives (Štetic, 2017). The assurance of a safe destination consists of a developmental strategy and it should cover the safety of its tourists, the employees in tourism activities, the security of the place to the point that it can be controlled by humans, and the general creation of a destination full of awareness and risk planning, able to reduce negative potential individual cases (Štetic, 2017). From a business aspect, one of the most important types of risks is the entrepreneurial risk or, in simple words, the risk that exists in the business environment. This is a sum of risk types including financial, competitive and reputational risk, and it requires all entrepreneurs to use their financial sense and business passion to predict their activity in the market so they can produce a realistic business plan which would be beneficial for future decisions (Ruan & Tian, 2019). The fact that there is continuous global market development and competition, characterized by technological and other innovative plans, enhances the doubts about the validity of the risk management process.

The validity of risk management that is established in a destination can be analysed if the given factor is tourists' revisiting intention. According to Hashim et al. (2020), the intention to revisit a destination is based on four components which are financial, social, physical reasons and time. Other theories, for example the one which has been conducted by Pennington-Gray and Schroeder (2013), mention that the validity of the risk management process is related to seven factors which are crime, diseases, weather, cultural barriers, political crises, equipment failure and physical reasons. Taking into consideration the factors mentioned above regarding the reliability of a destination, it can be said that they are related to tourists' age, gender, profession, education and family situation (travel with children) as those characteristics are important for the way of thinking and creating the final decision for visiting a destination, either for a first time or for revisiting (Karl, 2018).

As tourism is a sum of other individual industries, it is essential to understand all the unsuspected risk types for choosing the most appropriate risk management model for sufficient results. Due to the fact that the tourism industry creates a continuous competitive environment with the establishment of new services and new products, innovation risk is considered to be one of the most important risk types that exists within the industry. Innovation risk refers to change and the establishment of new services and ideas for creating a better environment for tourists and business (Williams et al., 2021). The more complex the innovation of the sector, the more severe its impacts will be (Merton, 2013).

Considering that operational risks, commercial risks and financial risks are classified under the risks of innovation, it is understood that risk estimation of an innovative activity influences to a degree the economic situation of a business, its internal operation (this may include meeting a quality and quantity target or the rescheduling of requirements) and its commercial success or failure, to the point of advertising successfully the business and attracting more customers (Nechaevet al., 2014). Businesses within the tourism industry have to deal with four main types of innovation and their hidden risks. Type one includes the excluded innovation activities which are almost the same with the activities that are measured by traditional indicators; for example, the development of a destination to be attractive by more tourists (Thomas, 2018). The second type of innovative risks is associated with new technologies, new business strategies and new business relationships with suppliers. An example is the replacement of expensive human labour with technological labour or the opportunity that tourists have to visit platforms where they can arrange their holiday by themselves (Hall & Williams, 2019). The third type refers to innovation which combines existing technologies and processes for offering hospitality and delivering customer service such as hotel functions and employees' categorization. The fourth type includes innovative plans that are not highly visible but have a huge impact on local economies; for example, the all-inclusive systems that have been established 20 years and their impact still on local economies. The ways that these four types of innovation are applied to the tourism industry can differ among destinations, businesses and countries and depends on the plans that are followed for the development of a tourism business in cooperation with local available activities (Hall & Williams, 2019).

Innovation contributes to new services development and assures growth and wealth sustainability for businesses, especially for those in the tourism industry, as they are the ones that offer their clients the opportunity to choose the products and the services themselves or what to invest their money on. For example, they can compare components among a huge range of destinations and take a final decision (Pikkemaat & Peters, 2006). On the one hand, innovation enhances the tourism industry through the creation of a smooth environment for the establishment of technological achievements and the adjustment to rapid changes that offer an extended variety of choices. On the other hand, innovation contributes to greater competitiveness that may cause big damages and losses to individual small and medium sized tourism businesses.

Innovation is characterized by the combination of known risks and uncertainties – the unknown risks. During the process of innovation, entrepreneurs are dealing with a range of unknown factors and their decisions have to be made under conditions of uncertainty (Williams et al., 2021). A study about the role of uncertainty in tourism innovation has been conducted by Verreynne et al. (2019) who recommend a model where tourism turnover connects with innovation diversity and business performance. The model is based on the notion that tourism turnover restricts the positive relationship between innovation diversity and business performance, while there is a positive indirect impact of uncertainty on performance through innovation diversity, depending on the size of turnover. The fact that innovative plans and recommended business risk management models have moderated the hospitality and tourism industry, demands the continuous improvement of strategies that businesses follow for dealing with unknown risks and potential negative results. Despite the fact that models which indicate ways to deal with crises on businesses' internal operations, marketing, organization, customer management and technological achievements have been discovered (Bilgihan & Nejad, 2015), there is still a lack of ways to deal with the establishment of new services in the tourism industry that will combine a smooth business adjustment to the new service and simultaneous improvement of the customer service quality.

Risk management in tourism industry in Greece

Despite the successful growth of the tourism industry in Greece, the country shows lack of risk management techniques as the field only started to develop very recently. There is no available literature sources to indicate steps that business should follow to deal with potential risks. Limited studies, such as the one that was conducted by Drogalas et al. (2017), show that the Greek perception for risk management is related to the internal audit function of a business, which evaluates the business's internal controls including corporate governance and accounting processes. The target of this kind of audit is compliance with regulations and the contribution to financial reporting and data collection (Gazdar & Cherif, 2014). Other studies have been provided regarding risk management in public health and such an example is the research by Gkogka (2011), who proved that the risk management model can analyse the impact of foodborne illness within businesses, aiming to a prioritization of food safety management. Additionally, there are studies about the establishment of risk management regarding regional storms and floods due to the weather conditions in Greece, and studies about the banking system in the country.

Taking into consideration the contribution of the tourism industry to Greece's GDP, the industry has the most significant role for national income. According to Statista (2020), the percentage of contribution of tourism to Greek GDP the last five years has increased from 19.2% to 21.2% and this reflects a 5% contribution to the total growth of the economy. As the statistical data illustrates, the importance of tourism as an economic activity for Greece creates a major need for sustainable tourist activity. According to Spenceley and Snyman (2017), sustainable tourism is based on criteria such as the socio-economic and environmental impact that the tourism activity has on the destination, tourists' loyalty to hotel chains and the innovative activities that tourists will be able to enjoy by visiting the destination. Factors that also affect sustainability are the financial system of a country that needs to be fair, and the proper leadership and management of the tourism industry that will enhance tourists' confidence to feel safe and visit the destination. It has been proven that experiences and sustainable tourism are interlinked, but further research on management will improve tourists' perceived experiences and will enhance each destination's position as a powerful one to survive in this competitive industry (Breiby et al., 2020).

Tourism is considered as one of the fastest growing sectors and contributes significantly to the development of a country by offering numerous benefits through the promotion of cultural awareness and destination image (Kyriakou, 2020). As it was mentioned earlier, tourism is a combination of industries that links tourism demand to supply and may be subject to a range of crises. A crisis can reduce the importance of tourism supply at the same time that tourism demand remains crucial for the development of a tourist destination (Mazzola et al., 2019). Being aware of the variety of strategies that may be applied for the success of a tourism business in a competitive environment, innovation has a key role to improve its position in global tourism competition through the establishment of attractive services and policies for tourists, as it is reflected in new ways of promoting and conducting business activities.

The Greek tourism industry is short on new services risk management processes and this may create problems for a country that is based mainly on the growth of its tourism sector. It is worth mentioning the innovative efforts that have been done for the development of the Greek tourism industry, which include participation in the online travel market (the travel industry is part of tourism), the use of social media as an marketing tool to build relationships with customers, the acceptance of world chain brand names to consolidate within the borders of the country and the cooperation with tour operators and travel agencies. Despite the fact that these components work positively for the development of the industry and the enhancement of economic growth, it is important to remember that Greece is a country apt to crisis, as it lacks a stable financial and political background. This makes the discovery of successful risk management processes necessary and demands concrete strategy for managing a turbulent environment caused by a crisis (Varelas & Apostolopoulos, 2020).

Based on the limited sources of literature regarding new service risk management or innovation management in Greek tourism, it would be beneficial if the steps of the risk management model are modified. From the potential establishment of a new service, the risk should be identified and the impacts that will occur in a long-term period should be examined, as the success or the failure of a new service will be proven within three years of its first appearance. The risk should be analysed to answer questions related to the necessary time the new service potentially takes to adjust to an existing industry with a given socio-economic background, the cost that will be needed for the full establishment, the ways that the new service will be advertised and attract tourists, the processes that employees may need for potential new training, and the background of the country where the idea comes from. Risk should be evaluated by taking into consideration the benefits or the losses and there should be an acceptable set of strategies to eliminate the potential damage or loss in the scenario of taking the wrong risk.

The Greek National Tourism Organisation recommended a marketing plan which aims to empower the tourist regions of Greece and recommends a variety of competitive holiday packages that combine famous destinations with emerging ones and offer to tourists a bigger range of choices and experiences. The Greek Ministry of Tourism prioritizes the development of new strategies through modifications in the collaboration of the public and private sectors, better practical training for people to boost their skills and employability and the partnership with other ministries such as the Ministry of Culture and Sports to create a healthier and more attractive environment for visitors. The discovery of services and products creates a continuous rising competitive environment and calls for businesses which operated within the tourism industry to face the new services as new perspectives for the improvement of the sector. Even if the process of establishing new services seems challenging, the percentage of revenue growth from the establishment of new products or new services ranges between 16% to 20% and this is vital for countries whose economy is based mainly on tourism.

Conclusions

New products and new services can be a powerful component for the growth of a business, for the extension of a sector and for the financial improvement of a country. Countries such as Greece that rely on tourism should accept the establishment of new services with willingness to create a more attractive environment for tourists. Taking advantage of the natural environments and the cultural heritage, Greece can be a representative example for a new tourism generation and could be the country with a key role for the creation of tourism dimensions. Without casting aside the traditional processes for conducting business, the country can employ a variety of strategies and be more familiar with the technological developments that, nowadays, affect most of the population.

The development of a new services risk management model which includes the steps that were mentioned above will avoid the creation of problems caused by the appearance of an innovation. The gap in the literature review about the new services in the tourism industry in Greece creates a fertile space for further research and the development of strategies. After the analysis of the steps, it is important to mention that Greece has the advantages of its location, climate and its culture to keep its tourism activity sustainable. Alternative types of tourism may be developed, cultural festivals with long term duration could be organized and the better use of new technological means for marketing can be encouraged for the overall improvement of the tourism industry. As innovation will be very productive for the enhancement of Greek tourism, the establishment of a risk management model focussed on new services is necessary for avoiding damages and losses and for the enhancement of a safe environment, attractive for tourists to visit.

In order for the Greek tourism to be developed and enhanced with ideas of new services, it is essential to cover the theoretical gap that exists from the first introduction of the risk management sector to recent studies that seem deficient for covering risks from the establishment of new ideas. Trying to balance a country with political and financial instability with tourism success is demanding, yet the analysis of strategies and the creation of more efficient business plans will allow taking risks and will protect destinations and business in case of unexpected results.

Highlights

- ☐ This chapter focuses on risk management sector and highlights its meaning as a recognized important sector for the global market.
- ☐ Through this report, it is pointed out the association of the risk management procedures with the tourism industry.
- ☐ This chapter direct attention to Greek tourism and the establishment of new services.
- ☐ This report gives information about risk management model and its adjustment to the Greek tourism industry.

Questions

1. Think about the meaning of new services in Greek tourism. Why is this concept challenging but useful? What is the reason for establishing new services within the tourism businesses in Greece and what is the meaning of new services?
2. It is important to think about tourism as a mixture of industries that have numerous risks. In this case, which are the potential unknown risks that may occur after taking the decision to establish a new service?
3. Which is the best available way to modify the risk management model to the new service risk management model and which are the relevant components that should be examined for the efficient use of the new model?

Reference List

Aven, T., (2017). How some types of risk assessments can support resilience analysis and management. *Reliability Engineering & System Safety*, 167, 536-543.

Bilgihan, A. & Nejad, M., (2015). Innovation in hospitality and tourism industries. *Journal of Hospitality and Tourism Technology*, 6(3), n.a.

Blumenthal, S., Williams, A. & Wood, H., (2018). *Assessing Risk*. Boca Raton, FL: Routledge.

Breiby, M.A., Duedahl, E., Øian, H. & Ericsson, B., (2020). Exploring sustainable experiences in tourism. *Scandinavian Journal of Hospitality and Tourism*, 20(4), 335-351.

Crockford, G.N., (1982). The bibliography and history of risk management: Some preliminary observations. *Geneva Papers on Risk and Insurance*, 169-179.

da Rocha, M.M. & da Silveira, M.A.T., (2021). Risk management in tourism: Analysis of destinations in Brazil and vulnerability to natural disasters. *Marketing and Tourism Review*. 6(1), 1-32.

Demir, E., Gozgor, G. & Paramati, S.R., (2019). Do geopolitical risks matter for inbound tourism? *Eurasian Business Review*, 9(2), 183-191.

Diebold, F., Doherty, N. & Herring, R., (2012). *The Known, The Unknown, And The Unknowable In Financial Risk Management*. Princeton, N.J.: Princeton University Press.

Dionne, G., (2013). Risk management: History, definition, and critique. *Risk Management and Insurance Review*, 16(2), 147-166.

Drogalas, G., Eleftheriadis, I., Pazarskis, M. & Anagnostopoulou, E., (2017). Perceptions about effective risk management. The crucial role of internal audit and management. Evidence from Greece. *Investment Management and Financial Innovations*, 14(4), 1-11.

Gazdar, K., & Cherif, M. (2014). The quality of institutions and financial development in meta countries: an empirical investigation. *Risk Governance & Control: Financial Markets and Institutions*, 4(4), 65-80

Gkogka, E., Reij, M., Havelaar, A., Zwietering, M. & Gorris, L., (2011). Risk-based estimate of effect of foodborne disease on public health, Greece. *Emerging Infectious Diseases*, 17(9), 1581-1590

Hall, C. M., & Williams, A. M. (2019). *Tourism and Innovation*. UK: Routledge.

Hashim, N.A.A.N., Mohamad, S.R., Abdullah, T., Awang, Z., Omar, R.N.R., Zain, E.N.M., Safri, F.H.M., Nawi, N.M.M., Yusoff, A.M. & Abdullah, F.A., (2020). Tourist risk perception and revisit intention: Examining the validity and reliability of the measuring instrument. *European Journal of Molecular and Clinical Medicine*. (in press)

Hopkin, P., (2018). *Fundamentals of Risk Management: Understanding, evaluating and implementing effective risk management*. London: Kogan Page Publishers.

Karl, M., (2018). Risk and uncertainty in travel decision-making: Tourist and destination perspective. *Journal of Travel Research*, 57(1), 129-146.

Kendrick, T., (2009). *Identifying and Managing Project Risk*. New York: American Management Association/AMACOM.

Kopp, E., Kaffenberger, L. & Jenkinson, N., (2017). *Cyber Risk, Market Failures, and Financial Stability*. International Monetary Fund.

Krzemień, A., Więckol-Ryk, A., Duda, A. & Koteras, A., (2013). Risk assessment of a post-combustion and amine-based CO2 capture ready process. *Journal of Sustainable Mining*, 12(4), 18-23.

Kunreuther, H., (2002). Risk analysis and risk management in an uncertain world 1. *Risk Analysis: An International Journal*, 22(4), 655-664.

Kyriakou, C., (2020). *Analyzing and comparing tourism demand in Greece and Cyprus forecasting scenarios for the future*. Dissertation. Open University of Cyprus.

Lemos, F., (2020). On the definition of risk. *Journal of Risk Management in Financial Institutions*, 13(3), 266-278.

Mazzola, F., Pizzuto, P. & Ruggieri, G., (2019). The role of tourism in island economic growth and resilience. *Journal of Economic Studies*. 46(7), 1418-1436.

Merton, R.C., (2013). Innovation risk. *Harvard Business Review*, 91(4), 48-56.

Nechaev, A.S., Antipina, O.V. & Prokopyeva, A.V., (2014). The risks of innovation activities in enterprises. *Life Science Journal*, 11(11), 574-575.

Pennington-Gray, L., & Schroeder, A. (2013). International tourist's perceptions of safety & security: the role of social media. *Matkailututkimus*, 9(1), 7-20

Pikkemaat, B. & Peters, M., (2006). Towards the measurement of innovation—A pilot study in the small and medium sized hotel industry. *Journal of Quality Assurance in Hospitality & Tourism*, 6(3-4), 89-112.

Ruan, X. & Tian, Y., (2019). The importance of risk and decision making in hospitality and tourism industry. *Frontiers in Educational Research*, 2(6).

Scaillet O., (2003). The origin and development of value at risk. Modern risk management: A History, *Risk Publication*, London, 151-158.

Seaman, J., (2020). PCI DSS. Castleford: Apress, Berkeley, CA,113-146.

Siegrist, M., (2021). Trust and risk perception: A critical review of the literature. *Risk Analysis*, 41(3), 480-490.

Soroka-Potrzebna, H., (2018). Risk identification as a basic stage of the project risk management. *European Journal of Service Management*, 27, 277-283.

Spenceley, A. & Snyman, S., (2017). Protected area tourism: Progress, innovation and sustainability, *Tourism and Hospitality Research*, 17(1), 3-7.

Statista. (2020). *Travel and tourism: economic impact by type Greece 2017*, Statista. https://www.statista.com/statistics/644590/travel-and-tourism-contribution-to-greece-gdp-by-type/.

Štetic, S., (2017). *Risks in Tourism*. Belgrade, Servia: Quaestus.

Thomas, T., (2018). *Inclusions and Exclusions of Social Tourism.* Malaysia: Taylor's Press.

Varelas, S. & Apostolopoulos, N., (2020). The implementation of strategic management in Greek hospitality businesses in times of crisis. *Sustainability,* 12(17), 7211.

Verreynne, M.L., Williams, A.M., Ritchie, B.W., Gronum, S. & Betts, K.S., (2019). Innovation diversity and uncertainty in small and medium sized tourism firms. *Tourism Management,* 72, 257-269.

Williams, A.M., Rodríguez Sánchez, I. & Škokić, V., (2021). Innovation, risk, and uncertainty: A study of tourism entrepreneurs. *Journal of Travel Research,* 60(2), 293-311.

Zou, Y. & Zhu, Y., (2020). *Handbook on Tourism and China,* UK: Edward Elgar.

C Conclusion

The Ancient Greek philosopher, Heraclitus, said "τὰ πάντα ῥεῖ καὶ οὐδὲν μένει" which roughly translates into *"everything changes and nothing stays"*. Along the same vein, Heraclitus shared another saying, stating that *"no man ever steps in the same river twice, for it's not the same river and he's not the same man"*. What these phrases summarise is that everything constantly changes and, in this change, both a positive and a negative aspect can be found. While pleasant moments pass so does hardship. And as the world around us changes, so do humans and, by extention, societies.

One industry that this doctrine applies to is tourism. The global tourism environment has been in constant change ever since the 1960s and the birth of modern tourism as we know it. There are innumerous publications supporting this argument, with scholars attempting to examine the way external and internal factors change existing processes, practices and structures in the tourism industry. Likewise, a plethora of studies have been devoted to the investigation of human behaviour in tourism highlighting that as structures, systems and environments change so do the tourist mindset and behaviour. In this pool of research, destination planners and business managers can find valuable insights which they need in order to make informed decisions about tourism development plans and strategy formulation. Whilst existing literature draws heavily on key trends, challenges and opportunities in tourism and hospitality, it has been less successful in offering a collective of new perspectives, challenges, and directions in global tourism and hospitality. This edited book has tried to fill this gap and incorporates a collection of chapters that discuss the changes emerging in global tourism as well as the effects that these might have not only on the industry but also on the global economy and society. In so doing, this book addresses the importance of key trends shaping the industry. while offering a better understanding of the rapidly shifting business environment characterising global tourism.

As a concluding remark of the *Tourism Dynamics: New perspectives and changing directions*, we provide a summary of the key issues and aspects raised in the book's chapters with the aim of offering suggestions and recommendations to destination planners and business managers as well as directions for further research to scholars of the travel, tourism and hospitality fields. As discussed earlier, tourism is characterised by a dynamic environment that is constantly changing and evolving. Existing knowledge about the tourism industry's operations and practices needs to be reconsidered as adaptations are required to meet the emerging needs and demands of tourists and deal with the ever-changing policies, structures and practices of the different sectors of the industry (e.g., travel, hospitality etc). Divided into three parts, this book sheds light on these important issues and aspects that arise due to emerging trends, changing patterns and developments experienced in global tourism.

The first part includes chapters that focus on the travel and tourism aspect, discussing tourist attitudes and experiences as well as providing a futuristic perspective on emerging trends in transport and tourism. Specifically, Chapter 1 explored the generation Ys' attitudes and behaviour towards engaging in volunteer tourism whilst backpacking. Chapter 2 examined experiential tourism in relation to improved holistic well-being within a protected natural area context. The third chapter explored the link between tourism and consumer neuroscience, discussing potential implications in terms of tourism planning and management. The fourth chapter discussed air connectivity in relation to leisure tourism, referring to the impact of the COVID-19 pandemic that has had significant effects on the aviation industry due to travel bans and tourism demand decline. Chapter 5 examined the munro-bagging activity as part of leisure tourism. These chapters provide significant insights into the rapidly changing environment of travel and leisure tourism, discussing evolving tourist needs, tourism forms and transport operations.

The second part of the book incorporates chapters that present destination related aspects and issues. For instance, Chapter 6 drew on destination community wellbeing to discuss community engagement in tourism using futures thinking techniques. Chapter 7 examined the phenomenon of overtourism, yielding significant insights on the management of different types of destinations particularly in post-COVID19 era. In Chapter 8 the importance of specific dark tourism motives (namely those relevant to war and terrorism) were explored in relation to dark tourism development whereas Chapter 9 drew from suppliers operating in a distressed destination to analyse government initiatives' role in tourism competitiveness. Chapter 10 discussed opportunities of using Extended Reality (XR) in tourism, providing contemporary examples of XR applications. In their totality, the chapters highlight the important effects of emerging trends such as technology and sustainability on destination development and management.

In the last and third part of the book, chapters examining emerging methodologies in hospitality research as well as educational aspects and risk management issues are included. Chapter 11, for example, provides an interesting comparison between linear and non-linear analysis whilst presenting complexity theory as an emerging and popular methodology in the hospitality research domain. In Chapter 12 the issues that face the academia providing hospitality education are addressed and practical recommendations are offered to education stakeholders in order to improve curricula and teaching and learning methods. Chapter 13 is the last chapter of this book and highlighted the importance of risk management in hospitality. As such, the chapters highlight the evolving patterns in hospitality which need to be considered by various stakeholder groups pertinent to global tourism and hospitality when dealing with policy formulation, management tactics and research.

Overall, *Tourism Dynamics: New perspectives and changing directions* provides conceptual and practical evidence which enables the adoption and implementation of appropriate management and developmental approaches in travel, tourism and hospitality by outlining the challenges facing the current global tourism industry as well as the opportunities emerging in the ever-changing environment in which the industry operates. Clearly, there is no standard formula for success as every destination is different, possessing specific contextual factors and characteristics. Nonetheless, destination planners, policymakers, researchers and students can benefit from a reading of the collection of key themes and issues presented in the chapters of this book and which bring about significant changes in the global tourism industry. Evidently, challenges and opportunities will continue to emerge in the dynamic tourism environment; yet, the improvement of management and planning strategies can only be achieved when understanding of these changes and trends is obtained by all the stakeholders involved in the industry.

Index

I

I

www.ingramcontent.com/pod-product-compliance
Ingram Content Group UK Ltd.
Pitfield, Milton Keynes, MK11 3LW, UK
UKHW021828100926
8556IPUK00003B/223

9 781911 635932